Human Ecology and the Development of Settlements

FRONTIERS IN HUMAN ECOLOGY

1974 • HUMAN ECOLOGY AND WORLD DEVELOPMENT
Edited by Anthony Vann and Paul Rogers

1976 • FUTURE RESOURCES AND WORLD DEVELOPMENT
Edited by Paul Rogers

1976 • HUMAN ECOLOGY AND THE DEVELOPMENT OF SETTLEMENTS
Edited by J. Owen Jones and Paul Rogers

Human Ecology and the Development of Settlements

Edited by

J. Owen Jones
Commonwealth Bureau of Agricultural Economics
Oxford, England

and

Paul Rogers
The Polytechnic
Huddersfield, England

PLENUM PRESS · NEW YORK AND LONDON

Library of Congress Cataloging in Publication Data

Main entry under title:

Human ecology and the development of settlements.

Includes index.

1. Human ecology–Addresses, essays, lectures. 2. Cities and towns–Addresses, essays, lectures. I. Jones, J. O. II. Rogers, Paul.

GF101. H85 301.31 76-10301

ISBN 978-1-4684-2267-2 ISBN 978-1-4684-2265-8 (eBook)

DOI 10.1007/978-1-4684-2265-8

The 1975 Commonwealth Human Ecology Council Lectures, given in London, 1974 and 1975

Softcover reprint of the hardcover 1st edition 1976

A Division of Plenum Publishing Corporation
227 West 17th Street, New York, N.Y. 10011

Foreword

This book is the result of a series of lectures organised by the Commonwealth Human Ecology Council as a prelude to the United Nations Conference on Human Settlements, the HABITAT conference, which will take place in Vancouver, Canada, in May and June 1976. The lectures were given in London, England, during 1974 and 1975, most of them sponsored jointly with the Royal Commonwealth Society.

Four years ago, the United Nations Organisation was preparing for a major international conference concerned with problems of the human environment. This was the UN Conference on the Human Environment that took place in Stockholm, Sweden, in June 1972. It was the culmination of a rising interest in the study of these problems in many countries of the world. The study of environmental problems relating to human settlements was on the agenda at Stockholm, but because of the great breadth of the subject of the conference they could not be considered in any depth. This will be rectified in Vancouver; and already the study of human settlements and their problems is the focus of an intensive programme of activities throughout the world in preparation for the HABITAT meeting.

The HABITAT conference is taking place at a time when it is recognised that human settlement problems are likely to increase greatly in severity in the remaining years of the twentieth century. We have entered a period of great uncertainty in matters of world development. The energy supply crisis has lead to the re-appraisal of existing strategies for the use of energy; problems of world food supply continue to cause grave concern; calls for a New International Economic Order grow louder month by month; at the same time we continue to be faced with intractable problems of world poverty and continuing population growth.

Over and above all these, the trend towards the concentration of human settlements in large complexes raises further problems. The urbanising process is world wide and its pace has been accelerating at an unprecedented rate. At the beginning of this century, barely 15% of the world's population of under 2000 million people lived in towns and cities of over 20,000 people, a total world urban population of 300 million. Now, three-quarters of the way through the century, the world's urban population has grown to more than six times that number, more than 35% of the world's population. The indications are that this will continue; so that by the end of the century, in only 25 years, there will be just under 3500 million people in towns and cities, more than half the world's population. In other words, the population of the world's towns and cities will have grown more than tenfold within the century.

The results of this rapid expansion are seen already in the appalling conditions existing in many cities and towns throughout the world. Shanty towns ring innumerable cities, basic public services cannot keep up with increasing demand and unemployment rises and adds to numerous other economic and social problems that collectively strain urban systems to the point of collapse.

But the HABITAT meeting is not just about the increasing problems of urban areas. It aims to study human settlements in their full complexity, in the recognition that attempts to solve problems of human settlements without reference to the total human environment, urban and rural, will be doomed to failure.

Coupled with this recognition is the realisation that human environment problems can rarely be solved unless we approach them in an inter-disciplinary way. Too often in the past, solutions to environmental problems have been attempted by the methods of a single discipline, and such efforts have created problems as great or greater than those they have solved. We are at last coming to recognise the importance of bringing together the major disciplines concerned with the study of the human environment. The Commonwealth Human Ecology Council is committed to this view.

The Council was established in 1969 with the aim of helping Commonwealth countries to approach by inter-disciplinary methods the solution of environmental problems associated with development. It has a wide range of activities including conferences, specialist symposia, publications and advisory work. Three major meetings on Development and Human Ecology have been held, in Malta in 1970, in Hong Kong in 1972 and in New Zealand in 1975. The Council, through its Communications Working Party, provides advice on educational aspects of human ecology and also serves as an information centre for human ecology research programmes.

As one of its contributions to the study of human settlements, and as a contribution to the preparation of the public for the HABITAT meeting, the Council organised a series of lectures which reviewed aspects of development in a number of Commonwealth countries, with emphasis on human settlement problems. For this purpose it was able to call on a wealth of experience from widely differing countries. The lectures published here represent the views of a variety of speakers all concerned with the inter-disciplinary study of human environment problems. They bring together the results of past and present work in five continents. They demonstrate the value of the kind of broadly based attack which is essential if we are to master the enormous problems that will confront us in the years ahead, resulting from the impact of modern scientific civilization upon the human environment.

HUGH W SPRINGER

Acknowledgements

We would like to thank Zena Daysh, Secretary-General of the Commonwealth Human Ecology Council, for her help, Greta Horne for assistance with the typing and the staff of the Royal Commonwealth Society for producing transcripts of the lectures.

Several of the contributors kindly provided us with edited versions of their lectures, making our job much easier, and we would like to thank the Hon Allan Isaacs and Dr Stephen Boyden for allowing us to include additional material pertinent to the subjects of their lectures.

Roy Baker of Plenum Press gave us much help and encouragement and we would like to offer particular thanks to Sue Bonham for undertaking the exacting task of preparing the final copy for publishing.

J Owen Jones

January 1976 Paul Rogers

Contents

INTRODUCTION

Ecology is concerned with the interrelations between living organisms and their environments, systematically exploring and determining the complex interacting systems upon which the organisms' survival depend. Human Ecology extends such studies to include the welfare and activities of man. The power and extent of human intervention in the environment is clearly of greater magnitude and complexity than that of other organisms. This power offers wide possibilities of using the environment favourably for human habitation: it also raises great dangers of environmental damage which could weaken and ultimately destroy the life support systems of human society.

Within ecology, it is possible to study both the broad principles of interacting and mutually dependent systems which one might call "macro-ecology", and the intimate aspects of adaptation and development of species and individuals to their physical and biological surroundings - or "micro-ecology".

Within "micro-ecology", and amongst the vital factors in survival and development are the habitats which organisms find, adapt, or create for themselves, using and changing their immediate environments to contribute to their security, livelihood and social exchange.

This suggested division between "macro" and "micro" aspects of human ecology are reflected in the balance of study between the two United Nations Conferences concerned with the human environment. Thus the 1972 Stockholm Conference concerned itself mainly with broad, global, "macro" problems: the projected Conference in Vancouver, 1976, will be concerned with the more intimate relations between man and his immediate environment in housing and settlements. This "micro" aspect has also been a predominant consideration of the Commonwealth Human Ecology Council (CHEC) over the past two years, resulting in the series of lectures recorded in this volume, and the statements and recommendations of the Commonwealth South Pacific Regional Conference on Human Settlements held in Auckland in November 1975, under the joint auspices of the City of Auckland, the University of Auckland and the Commonwealth Human Ecology Council. The statements and recommendations form a fitting focus for this collection of papers:

A. RIGHTS

1. Each individual person has the right to an effective habitation in which to achieve adequate shelter, health and social and economic needs and that this right must be set with the rights of life, liberty and work as fundamental ends of society; and
2. each individual has a right to demand that his human settlement needs will be regarded as a matter of priority in political decisions.

B. POLICY AND ACTION

Human Settlements - a complex issue

The Conference stressed the complex nature of human ecological issues, and warned that there were no simple solutions to the achievement of acceptable human settlements.

Urban-rural interaction

The importance of the interrelationship between settlements and adjacent rural areas was stressed and any good human settlement policy must aim to achieve a proper balance between the two.

Quality and Scale

The quality of the environment must be a prime consideration in settlement policy. The scale of settlements must be considered in relation to this policy.

Realism and attainable goals

Many more people will benefit if planning is carried out with realism and with goals that are achievable.

Public Participation

The need for more and better communication between planners, policy makers and committees, and the need for grassroots participation is paramount.

Self-reliance

Research and aid are recognised as important but only in engendering the spirit of helping local people to self-reliance.

Education

Education at all levels and ages is essential to sensitise people to the needs of human settlements and to contribute to greater social cohesion. The importance of active participation of the young is particularly stressed.

An ongoing process

All settlements policies must be seen as a learning experience with a continuing adjustment of policies in the light of experience.

Communication

Effective communication between all those involved in the creation of human settlements is essential. It should be a continuing, ongoing process, and should include information on mistakes as well as successes.

Land

Accessibility to land is a fundamental right of every human being. Sensitivity to fellow humans is as important as wise land use.

Cultural and Site characteristics

The characteristics of the site and the cultural aspirations of the people concerned with the creation of a settlement are of the highest priority.

Local building materials

There is a need for a careful evaluation of materials used in building. It was decided that immediate improvements in settlements would follow from the use of local building materials.

What are the minimum requirements of an effective habitation? Clearly these include protection from the elements, and the possibility of maintaining a tolerable internal temperature. The habitation should provide privacy and security for the individual and family, with adequate access to water, sanitation and cooking facilities. Beyond this, it needs to afford the physical basis of a home - including means for social, creative and recreational activity for both adults and children. Size, design and construction materials should be related to the economic situation of the communities concerned. There are abundant illustrations that this may not be a severe constraint when using local labour and materials, and when house-building is an integral part of a local development project. The Indian National Buildings Organization, for example, is concentrating on designs of rural houses costing between £30 and £90, drawing from the rich experience and traditions in the rural areas where the use of local materials and skills can result in practical housing or considerable aesthetic charm, and adding modern technology and knowledge of health requirements. The development of the Sontal-style housing in NE India provides a valuable illustration.

In a following paper, Dr Obeng describes a rural re-housing project in Ghana where individual dwellings cost up to £330 - again modest by Western standards, but using sophisticated and locally unfamiliar materials such as concrete and aluminium.

Necessarily the structure of human settlements reflects and influences the structure of society, and most provide for the needs of the individual, the family, and the larger groups which are interdependent in their economic and social functions. The habitation is a unit within the wider community in which specialisation, interdependence and mutual interests provide the setting for employment, creative activity and social expression. Thus beyond the individual dwellings it is necessary to consider their special inter-relations, and connections with community services and focal points such as shops, public offices, community centres, places of worship, and areas and buildings for education, recreation and entertainment, the whole complex being related to the basic productive work of the area.

Many modern urban developments, particularly those

concentrated in tower blocks, have signally failed to provide the basis of adequate settlements in spite of incorporation of highly sophisticated technology and great expenditure of resources.

Following consideration of "micro" aspects, it is necessary to relate to the "macro" situation, for the increase in numbers and areas of human settlements necessarily affects the global environment in various ways. These aspects are closely related to the size, rate of increase, and distribution of human population. Amongst the disturbing trends mentioned in the following papers, not only is world population likely to double itself by the beginning of the next century, but over 50 per cent could by then be living in urban areas unless there are significant changes in direction: Professor Gertler mentions that Canada is moving towards having over half its population concentrated at three points "if we let things ride". Looking further ahead a world-famous regional planner, Professor Constantinos Doxiades, has projected a stable world population of between 15,000 and 19,000 millions by the mid 21st century, occupying a virtually continuous conurbation, or "oecumenopolis" covering much of the world's habitable land surface.

It is questionable whether even such a rate of urbanization as is quoted by Professor Gertler, and which is almost upon us, is desirable, or even possible without severe human deprivation. Most people at present live in the rural areas; and in the low income countries where populations are increasing most rapidly, the proportion is typically 70-80 per cent. In the case of India, for example, the rural population of some 450 millions is increasing by about 9 millions each year. It is hardly conceivable that India could extend urban areas to provide adequate living and working facilities for this additional number each year, bearing in mind the present situation in her cities. It is worth pausing to consider what this would involve in terms of factories, power, equipment, training, raw materials and markets, in addition to housing and adequate living facilities with all the essential urban infrastructure and services. But even if this could be achieved it would do nothing to reduce the present unemployment, underemployment and inadequate living conditions in both town and country.

In other developing countries, absolute numbers may be less but proportionate increases are similar or even greater. In Bangladesh, for example, ninety per cent of the population lives in the rural areas, while the ratio or increase in population approaches 3.5 per cent per annum. In most cases, a far more desirable course than urbanization would therefore appear to be decentralization, with the expansion and creation of smaller settlements in the rural areas which would provide from the rural base and in serving the rural communities, the necessary variety of life and occupation to make these areas attractive to people of various aptitudes and backgrounds, checking and perhaps in time reversing the present flow to the already over-crowded urban centres. A technological factor aiding decentralization is the rapid development of communication facilities. The development of transport in the nineteenth century resulted in people and materials being brought together in steadily increasing concentrations: the development of communications in the twentieth century allows people to work together and, where appropriate, to use centralized facilities such as computers and data bases without the need to be physically in the same place.

A further ecological factor is the relation of settlements to energy requirements. Present trends, if continued, would soon lead to a rate of energy consumption hardly supportable continuously from known resources and technology, and which would be likely to add the environmental hazards of radioactivity and heat pollution. The United States of America already consumes energy at a rate of about 13 tons of coal equivalent per person per year. World energy consumption is at present doubling every 15 years. Smaller, decentralized settlements, designed to make best use of local renewable and largely non-polluting sources of energy including hydro, solar and wind energy and biological sources, offer a more attractive prospect in which material and energy consumption are maintained at a sustainable level adequate for human needs, and further growth is sought in non-material services and achievements.

Various ecological problems of human settlements drawn from widely differing situations are unfolded in the following pages, and aspects of themes developed are recorded by Professor Johnson-Marshall in the concluding lecture. He emphasises that planning for human settlements should be based on human environmental needs. It is a sobering reflection that after many years of active urban, rural, and regional planning in most countries, these needs have still to be defined. Professor Johnson-Marshall suggests that a beginning could be made in considering basic environmental standards such as minimum spaces for living, working and recreation. It may be that these standards themselves will vary according to the setting of the individual housing units. One could be happy in a small unit where there is enjoyable space outside, but greater concentration of habitations may necessitate greater space allowances for the individual units. A further step would be to incorporate the derived standards in designs which minimize costs in terms of energy and materials. Consideration should also be given to "social costs" - those costs not incorporated in particular structures but which nevertheless may have to be borne by the community, for example, in terms of pollution or scenic disfigurement. The Commonwealth, through CHEC, could well take the initiative in such investigations. It is clear from the following papers that the Commonwealth embraces the whole spectrum of human habitat situations in all possible environments, but at the same time benefits from a range of unifying factors including common language and traditions, mutual sympathy and understanding, and wide experience of collaboration. Great possibilities are offered for defining human settlement needs, and translating these definitions into reality in providing suitable forms of human habitation within and around which people may achieve their full potential.

J OWEN JONES

HUMAN ECOLOGY, SOCIETY AND COMMUNICATION

JOHN ROBERTS

I am, perhaps unfortunately, not going to reveal anything particularly new or develop any new thesis, because I am not really concerned with the margins of my particular discipline. I am concerned, in fact, to consolidate a movement that I think is extremely valuable and which will need the help of all those people, who are not normally accessible to the aspirations and the needs of the sort of people that I want to talk about. I believe this is extremely important to the Commonwealth Human Ecology Council and what I want to argue is that the movement for the care of human ecology must be rooted in the political realities of community if it is to have a practical outcome. And that broadly, and rather simply, is the purpose of this disquisition.

I'd like to introduce it by saying that I want to talk about typologies. Do not be surprised by this particular word. A sociologist is a man constantly amazed by the obvious, and I am by way of being a sort of one. Well, typology is only a "high class" way of saying I want to talk about two kinds of societies - the great society and the small society. And I mean by this the great society which has the guns - that is - the one with coercive authority within; for our purposes, a nation State, since a nation State has become the basic unit of political action. And the small society is the one that works for the most part, industrially and not with much hope, to supply the managers of the great society with their capacity to maintain order.

Those of you who are concerned with political philosophy will know that the oldest and perhaps the most famous analysis of this functional division will be found in Plato's 'Republic'. One of the reasons why that irritatingly tenacious polemic retains its fascination for people like me, who are bothered by the phenomenon of power, lies precisely in this understanding of the fundamental dichotomy in the nature of politics. Now the great society has changed its form throughout history to suit whatever was the prevailing technology of power. If horses were needed to maintain power then the great society got a monopoly on the horses. If railways were needed to maintain power then the great society had a monopoly over the use of railways. And the one thing that remains constant is the necessity to secure a large share of the social resources available within society to maintain the authority of the great society. Marx, who was I think, a modern Platonist - and it's just as well that the old gentleman from Highbury is not alive because he would certainly reject this definition - was surely right about this; that the modern forms of great society government and private industrial society bureaucracy rest upon the mature physical and social technocracy and draw their strength from their control of the means of production.

The small society, on the other hand, unlike the great society, tends to change little. It is preoccupied with work, with shelter, with food, with survival and the maintenance of a close community relationship, with the family, and the nature of its political presence in the body politic does not really alter materially with the changes in other social organisations. In one sense we can see the process of modern political evolution as an attempt to bring the small society and the great society into an effective relationship. The history of democracy, which, you will recall, was that form of constitution that Plato found most dangerously perverted, is littered with attempts to equate the interests of the great and the small societies by giving to the individual an element of control over the access to power, and the day before I arrived in Britain you were exercising that form of control, with what satisfactions I leave yourself and your conscience to meditate.

There's no doubt that this has been successful in mediating the use of power over the small society and only the ideologically blind can deny that the quality of human settlements, the services available to communities and the chance for individuals to enjoy the powerful support of the great society, have been improved by the responsibility generated through systems of democracy. And it's pleasant here to say that it is my considered belief that this is nowhere more true than in those countries that have evolved their political institutions under the influence of the British experience.

Now, as the High Commissioner knows, I am the son of an Irish Socialist and a Scot whose background runs among those poor exploited military servants who were pressed into defending the Empire over the Berlin plains and under scorching suns, and my own admiration for that tradition is not by any means automatically acquired. Yet if I were forced into a corner and asked which great political act constituted the greatest political advance in the people's struggles for political control of their destiny, it would not be the Declaration of the Rights of Man, it would not be the Reform Act, nor would it be the Communist Manifesto. It would be the Municipal Corporations Act in this country in the 19th century, for it seems to me that at the heart of this matter one will have to deal with effective political representation - those levels that touch the small society in all its aspects.

I do not claim to be an expert in comparative government yet in the years I have been studying politics I can find no other system in which a just balance of power between the great and the small societies has been so nearly achieved as in this country and in those States which have modelled their politics on British experience. In that respect, for men and women in their social context, and in the delivery of services honestly and painstakingly conceived for the individual's benefit, I do believe that the Commonwealth has the experience to claim a hearing and the base from which to lead a struggle for a society rooted in the care for human ecology.

That the network of relationships of man as a social and biological phenomenon to his environment, which I take to be the field of human ecology, is threatened in a deeply serious manner may be I think, taken as read; perhaps it is a little more controversial to suggest that this is to a large extent the result of the depredations made upon our social and physical resources by the developing great society of mass government and mass economics. Perhaps I could quote from a Commonwealth Human Ecology Council publication, because in a book published as a result of a symposium at Huddersfield - and this is in a book edited by Anthony Vann and Paul Rogers - Peter Kenyon suggests that the current world disorder is dominated by a configuration of the phenomena - Science, the Market and the Nation State - an unholy Trinity. The characteristics of these structures is that they must acquire what I propose to call system control of resources. As J K Galbraith has pointed out, in the new industrial State these institutions are necessarily functional planners. That is, to ensure the continuity of the goal set by the functional nature of their power they must forecast their resources and impose their will upon those who supply labour and raw materials. Often enough they are isolated from the claims of the small societies which are necessarily whole and indivisible and indeed they can be plain hostile to them. One of the sad facts today, is that to work the complex mechanism of great systems one has to be already rich, educated and powerful.

Barbara Ward, in a brilliant analysis of the current economic crisis which is published in the United Nations publication "Development Forum", points out that in Brazil, the phenomenal increase in wealth in that country accrued very largely to the most prosperous fifteen per cent of the people. The masses were not much better off and the ten per cent at the bottom were actually poorer. She goes on to a passage which I think could be taken as a classic statement of the great society effect upon human ecology in most developing countries:- Investment was moving into the modern urban sector, blowing up cities to make offices, attracting a rising stream of immigrants, yet exposing them to the risk of high unemployment. Meanwhile on the farms too, employment was ill-sustained, either as a result of little investment or preoccupation with the highly capitalised energy intensive farming of the green revolution type. In both cases the exodus from the land increased. By the end of the sixties the 'marginal men' could be counted by the million on run-down farms and shanty towns almost as divorced from effective market demand as were the 19th century Irish immigrants of the hungry forties and the despairing weavers starving all over Europe as the machines moved in. With the new poor, as with the old, an almost complete redistribution lay at the root of their inadequate claims on the market.

Barbara Ward suggested that these vast economic

processes created a dilemma of survival and that would stand as a good statement of one conventional view of the problem the world faces. It is the question whether the international economic system, proposing to itself the patterns of technology, investment and energy use of the last two or three decades, can find the resources to satisfy the demand of what may be seven billion people thirty years hence, and to do so in such a way that the basic environmental balance of the planet is not fatally disrupted.

Now when someone in my trade is made critically aware of those processes that we call the environmental crises, the mind automatically throws up two questions. Why have the long established environmental controls failed and what must be done to reform them? And I have to admit that a good deal of my time has been spent in the last few years on what I now believe is a fruitless quest for specific environmental standards universally applied. When one considers that a century has passed since Mayhew and Booth revealed the frightful effect of ecological deterioration, not far from this lecture hall, in human communities and that the great exemplar of humane legislation, the Health Act, goes back to 1877, when Ebenezer Howard had provided a vigorous opening shot in the campaign for life enhancing urban environments, and that men like Brunel had solved the conceptual problems of providing mobility in a mass society, it becomes clear that it was not so much the default of specific policies, but the means of applying and developing community power itself that demanded reconsideration.

Now I cannot say I reached this conclusion willingly. Like most academics I have an exaggerated respect for systems, even though I do not understand them very well. I bow in admiration before statisticians, mathematicians, systems theorists and other marvellous people who speak those wonderful languages to which I only wish I had access. The theories that treat organisations as organic structures accessible to cybernetic analogies were powerfully attractive as models of institutional control. Theories dependent upon the seductive symmetry of model building where the sweating reality of human conflicts were dry-cleaned by the simulation process seemed to be the lever to move the modern world.

Sad experience in helping to set up a Central Plan

Bureau in my own country and observation of the process elsewhere, has convinced me that these institutions are necessary but they have very strict limits of utility. Now the orthodoxy of people in my position is to espouse reform, reinforced by one of the varieties of social democractic politics on the belief that political institutions must be used deliberately and systematically to order private relationships in the pursuit of the public interest to the achievement of equality by gradualist reform. Now I am forced to the conclusion that this philosophy, which in fact I still espouse and will, out of filial piety if nothing else, to the end of my days, has been exported with tragic results to developing countries without modification and, with its centralising system-building instinct, has undermined traditional institutions more sensitive to the dynamics of human ecology, but has not developed structures to replace them.

I wonder if I could, just briefly, refer to a personal experience. There is a town close to my own city, called Porirua. And there is also, as well as the institution of Porirua, a large government housing estate established east of the commuter railway line. In the west live the middle class, engaged in a glacial slow motion combat with their mortgagees. I have recently written on the ward system in local government - that is, area representation to a central council and therefore was asked for that reason to join an otherwise very distinguished panel consisting of the then Mayor of Wellington, who is the world's living expert on the subject and asserts that he is the longest serving Mayor of a capital city in human history, and I certainly would never dare dispute the fact with him; New Zealand's foremost social historian, and my own senior Professor, who is also an experienced local government politician.

The panel harangued the Mayor and Council and citizens of Porirua on the iniquities of this ward system because it breaks this centralising planning function. We demonstrated the narrow localism upon which it is based and the way it offends the broad principle of the public good. We showed how the untidy demands of the people would insidiously rule the minds of the councillors and the long view would be lost in the myopia of particularism. If I may say so we were very eloquent and full of wise words, very ingenious and quite patronising. And in the end the man who proposed the enquiry thanked us very much, very humbly, for setting him straight and pointed out that it was only because on the council of twelve, eleven councillors lived to the west of the railway line and one lived to the east. Still, he could see there was nothing in comparison with the awful disadvantages of the ward system.

Well, I have wrestled with my conscience on many occasions since then and I always lose. The effect of reform has been to disenfranchise the people that the reform has sought to help. The political system has become too complex, too large, too remote and too powerful for the sort of people who live in Porirua to move. And for most people government is no longer as it should be - part of the ecology to be easily and effectively influenced. It is too often an external and alien constraint upon their lives. One could suggest perhaps that this has always been the case, but it is not altogether true. Even in a society of slaves, government was directly visible and subject to influence. It has taken modern benevolence to create a structure - a system structure - in which there is no really visible source of control accessible to the government or in so direct a relationship that at least some genuine human inter-action occurs. The modern housing estate, whether it be Porirua or Peterloo, is the more dreadful, because like other ecological disasters, it is the product of the most benign intentions. Like the elimination of malaria in Ceylon, which has increased dreadful problems of over-population. Like the vast want-creation and satisfaction process of a modern junk civilisation where there is no desire to subject anyone to the treadmill of waste making. I suggest at base that these great unintended consequences are the product of massive functional systems that find it difficult to respond to and to question, and from that to alter and reverse, fundamental power decisions.

Now I am going to talk about functionalism, and I am afraid I must claim the privilege that was given by Lewis Carroll to Humpty Dumpty. "A word means what I choose it to mean", and I am going to use the word 'functionalism' in a way which will horrify any sociologists here, and I apologise to them in advance. There are no effective built-in, human-scaled, need filters that can take account of the individual in his complex ecological setting in this functional dilemma.

As mass societies have developed they have been locked into an ever increasing number of large decision-making systems concerned to integrate individuals as participants in particular functions. The great bureaucracies are the characteristic mechanisms of these systems, articulating power by the control of resources on a quantitatively huge scale, yet a functionally limited one. Among the earliest examples were the great mass conscript armies developed in Europe in the 19th century, and they were followed, indeed they were made truly effective by the transport structures of the railway age, when the relentless pressures on the individual as a commuter began. In the same era we saw the growth of mass educational programmes disciplined to standard curricula, the standardisation of community services for the delivery of health care, supply of water and drainage, energy, housing, roads, consumer goods and so on.

This silent revolution in human organisation made mass communication possible, provided the infra-structure for a vast increase in productive capacity and made practicable the tremendous urbanisation that we are now familiar with, which surely is the greatest identifiable change in human ecology since the emergence of Homo sapiens. With the exceptions of the role of population increase which we now know was disastrous in its implications for the global environment, and of course, the implications of mass warfare, these functionally specific technologically advanced systems have been generally favourable. Where they have been most vigorously developed people are better fed, better housed, work less, have better health and education, enjoy a more diverse recreation than in those places where they are as yet rudimentary, and it is no wonder that developed societies have become the models for the under-developed. As fast as they can they are concentrating population and evolving systems of massive resource concentrations.

We are all worried and puzzled by their lack of success on occasions, but no one has doubted that the same process of industrialisation, urbanisation and bureaucratisation, is essential if there is to be any global equalisation of living standards. And the cost in this process has been the small independent community providing an ecological setting for the human individual. In a structure in which individual family, clan and tribe merged into each other and in that inter-action provided work, food, housing, energy, a social life bounded by the intimate realities of the environment had to be accepted, but at least its consequences could be integrated within a pattern that the individual could understand. When this pattern comes up against the functionally specific urban existence of advanced societies there are inevitably great strains. Every society experiences them. The absorption of members of former slave societies into Detroit, the Pakistani Muslims into Birmingham, or Melanesian tribesmen into Port Moresby, is never anything other than painful.

Let me provide one example whose terror I think, lies in its very banality. Supposing a Tuppaloan family comes to my country - to New Zealand, and at the rate they are coming the Tuppalo Islands will be de-populated within a very few years. Now his life has been spent on an atoll where fishing and the preparation of copra are virtually the only occupations. Survival is not simple. Occasionally the Tuppaloan must lash himself and his family to a tree until the hurricane passes. Only with the closest and most involved ties of mutual aid reinforced by ceremonials of a most compelling nature can such a society survive. Well, naturally our Tuppaloan is poor. Capital formation on a non-cultural atoll is a difficult business. Unskilled in the systems of modern, highly differentiated employment structures, and dependent because of language and social adaptation, he has problems which make it difficult for him to cope independently with the society he has chosen to live in. Naturally, in such circumstances, the Tuppaloan attracts the benevolence of the State, which supplies him with perhaps a cheap house, probably in the Porirua housing estate, helps him get a job, and probably gives some social counselling from a white social worker.

Now, suppose a cousin arrives from the Tuppalos. The whole of our hero's traditions and culture require a big welcome to be given to his kin. Perhaps he thinks he should acquire a pig and fatten it up; feeling that his neighbour will understand the exigencies of hospitality, he then proceeds to slaughter it in his back garden with a slasher. Now those of you who have slaughtered a pig with a slasher in your back garden will know that as an occasion it is rather fraught. A slasher by the way is a very handy implement. You can build a road or a house with it, or clear the bush; and a friend of mine used to eat his dinner from one. And the neighbours complain when the pig dies noisily, and up until that time it is all innocent comedy, and then the police arrive. Then our Tuppaloan, offended, frightened, frustrated, takes to the policeman and, being a tough man as most Polynesians are, beats him up. And that is where the tragedy sets in. The New Zealand prison population has a much higher proportion of Polynesians in relation to total population than it has of European prisoners. People with all apparent political rights, indeed with some official preferment, are in fact unable to make their wants known because they are unable to manipulate systems. The story is depressingly familiar and our Tuppaloan is only an atom in the urbanising process going on everywhere in the world.

Now it would be easy to reject the great society out of hand and turn to the small as a sort of embattled last ditch stand to save the world, and there are some movements which move in that direction. But it would also be nonsense. In the first place the great systems are absolutely essential to maintain what we have created; we cannot do without them. In the second place they are often benevolent in their effect upon the quality of individual life. The problem is not fundamentally one of conflict between the great and small societies, but rather ensuring that where they penetrate each other the result is humane, life-enhancing and enlarging to the individuals who live in the complex social setting. It is my contention and the theme of this lecture that to ensure this result we must strengthen the political weight of the small society. In this the human ecologist must play a vital role because, and this is the supporting contention, he alone has the information and the intellectual discipline that can provide the small society with a credible statement of need and with an understanding of the ways in which the great society may be manipulated, and therefore he is an essential element in the process of securing support for community structures. But you must remember that human ecologists are expensive commodities. Biologists, doctors, planners, dieticians, architects, sociologists - they do not come cheap. Therefore they are generally the servants of great societies that can afford them.

The next phase in my argument is that in some way, partly as a result of the pressure of the sociologists themselves, the great societies must be forced to disgorge their resources to ensure that the small society secures the advice and assistance of those trained in the care of human ecology. Let me again, if I can, relate a small personal experience. There is in my town an area where for the most part people do not like living. It is poor, by and large run down, and it looks as though people do not care for that environment. Well, the Council, under a legislation that had been provided by the great society - a typical way of destroying the existing community and re-building without thinking about the effects upon the existing ecology - decided they would put in what is called a comprehensive urban re-development area. The Council went ahead with this but they overlooked one point. At either end of this district was (a) a Polytechnic and (b) the University, and some of the people - social scientists - who lived in the area and worked at the University and Polytechnic. Now they did not represent the local inhabitants, who did that for themselves, but what they did do was supply the local inhabitants with a back-up which showed quite clearly that the information upon which the Council had based its scheme was totally incorrect, and did not take into account the complexities of the local community. For example, in this area there was a colony of transvestites - rather pleasant people - and their Sunday morning basket ball game at the local school was a major attraction of the region. Now what would have happened if the Council's plan had gone through and they had put in these single unit, large, high-rise housing construction plans that they had proposed? A gentleman by the name of Big Boobs Mary, who is in fact, the leader of the transvestite colony, had asked the local housing council officer for a flat. Just what chance would he have? That was a question I had to ask because he was very helpful and very intelligent and very sensitive to the needs of the community when we were developing the submission.

Now by accident, this seemed to me to indicate the function of human ecologists in this field. To back-up, to investigate, to provide information, to systematise, but not to dominate the small society. And the conclusion I drew in this specific field is that in the whole process of community planning the evocation of a real political skill at the community level is the task of those who feel dedicated to the objective of improving the quality of life. I have been interested for some years in the reform of local government and we are at the moment, under the influence of the great movements that are occurring all over the world in this field, notably of course, in Scotland and in England under the influence of your two great Royal Commission Reports, in the structure of local government.

We now have a Bill that will re-structure local government, and one of the provisions is for community councils, neighbourhood organisations. I am hopeful that in fact we can provide the basic level of ecological services through the community councils, and I believe that that should be the objective, whether in developing or in developed countries. One of the problems of course is that the difficulties faced by my country, which is pretty wealthy and egalitarian, look so utterly unimportant against the problems of say a Bangladeshi leaving the delta to live as a marginal man on the outskirts of Dacca, but I do believe that this is one area where the experience of developed countries has a direct spill-over for the developing countries, and that a concentration in this field of study, based upon studies like 'People and Planning', which tries to integrate the services for human ecology at the level at which they can be understood, the level of the small community, can be a major element in ensuring that the balance of political power between the small and the great societies can actually contribute to the sum of human happiness.

Some, of course, of the solutions that have been adopted by developed societies are dead failures. One can see them around us everywhere. They, equally, can feed in to the understanding of these problems. The identity and definition of these problems is one of the most important tasks it seems to me, that human ecology has before it, and I hope that the Commonwealth Human Ecology Council will make it, in the next year as we move towards the Conference on Human Settlements, a major element in seeking to define the contribution that professional ecologists can make to the struggle for small societies to maintain an effective and developing place in the massive drive towards functional systems that characterises our era. I do not want to be romantic about it. I have got no romantic views about small communities. They can be instruments for suspicion, for small-mindedness and I believe that they have to be controlled by general structural legislation that forces upon them the pursuit of the public good, but I do not believe that is an impossible aim.

I also believe we must delegate political power. Enrique Penalosa, the Secretary-General of Habitat, has said this about the conference. The planning, construction and management of settlements, whether the rural village or major city, requires many hands and skills - the representatives of the Ministries of Housing, Planning, Urban Affairs, Public Works and Finance should be joined by national and local political leaders. Equally vital is the participation of the appropriate professions as well as representation of civic organisations dedicated to various concerns, labour organisations and so on. The conference will cut across a full range of human settlement and environmental issues, such as housing, air, water, waste disposal and transport, as well as individual and collective needs for cultural expression and participation. It will study and compare national development strategies and planning and management systems and community technologies. Now there is one thing I feel that we must be clear about, because although I believe that physical planning has an enormous contribution to make, in itself it is not the total. We must remember that human ecology takes the totality of man's relationship to his environment, and physical planning is only part. And one hopes that this conference will see a full range of people dedicated to the restoration of community systems.

I believe that there is much to learn from the anthropologists, who can tell us how traditional communities have managed this. In my own country there is a tradition of meeting both for ceremonial and collective decision purposes in Maori communities. It takes place in no more than an open space called the Mari in the village where the community leaders, plus the rest of the people in the social unit, meet to discuss the problems that are affecting them and meet also for the ceremonial process of greeting visitors. There has been a movement, of which at first I was extremely suspicious, called the establishment of the urban multi-racial Mari. Just before I came away I went to a Mari in a local community. It was not a formal political association. People - anybody - can get up and speak. There are recognised processes by which people are entitled to the attention of the Mari, but if you go through these ceremonies in a small rationing device, then you get a hearing. I was amazed at the vitality and the meaning that had emerged. And there is a purely - almost a stone age - tribal device which I think can clearly be relevant to the problems of a large urban complex industrial culture.

I think there are many other examples of the need to preserve and to re-work traditional systems of community inter-action for the purpose of protecting the small society against the depredations of the great. I believe that, in fact, the planning community generally, losing confidence in great systems has turned its attention to the creation of effective community environments. Provided they take into account the necessity to incorporate in their considerations the need for a political identity then this, it seems to me, will be a real advance in the business of ensuring effective human settlement in the future.

There is I think, a difficulty here, because we must merge disciplines and accept a willingness to evolve while preserving a fundamental respect for the privacy of the individual in all the social processes. But in the world of international exchange and mutual support, are there any other groups than the Commonwealth that offer a better chance to advance the quality of social control, and to put the care for human ecology at the heart of the political process? Here in the city that gave a painful birth to the Commonwealth itself, and where the idea of responsible government was equally painfully invented, I would feel that one's hope in human affairs had indeed sunk low if I could not ask for a critical and creative lead from the Commonwealth to restore and improve the quality of the human condition.

MAN-MADE LAKES AND PROBLEMS OF HUMAN SETTLEMENT IN AFRICA

LETITIA OBENG

Throughout the world today there is an urgent attempt to attain and maintain a standard of life which is satisfactory and acceptable to us as human beings. For a greater part of the world, and certainly for the part of the world to which I belong, this means simply sustained availability of good food, good health and good shelter. And much national effort and wealth are expended basically on these items, and particularly for good shelter. Provision of satisfactory human settlements is a problem for all countries as it constitutes a pivot around which community structures and therefore social, economic and cultural systems are formed. We of the non-industrialised developing world depend very heavily on our natural resources for the attainment of our type of life - the kind that we want - and therefore we undertake major development projects which are dependent on our land and water resources. We are therefore faced with ecological disturbances which accompany these projects and this is particularly true of river basin development.

Natural water resources such as rivers have always been unreliable, and sometimes quite unpredictable in their availability, and for centuries man has sought to accumulate, conserve and utilise water resources in various ways. And since our technology has permitted us, we have also in our time indulged in the modification of rivers in various ways to serve many purposes. Repeatedly and especially in our search for energy we have dammed rivers of varying sizes all over the world to obtain electric power. Since the later 1950s a number of Africa's large rivers have been dammed in major river basin projects, which have attracted attention widely because of the adverse ecological impact which they have had.

The inundation of large areas, we have to admit, has a sort of finality about it. A range of terrestrial ecosystems are destroyed. The resulting aquatic ecosystem may bring additional problems into the basin. The soil, the water flow, vegetation, animal community and even the underground water stability and climate of the basin may be changed. People are invariably affected when they are displaced by the project. Directly or indirectly, river basin development projects have repercussions on human ecology. The impact is usually very complex and it threatens the economic stability of the people as various resources are irreversibly destroyed by the flooding of the basin. There are grave social disruptions and a tendency towards cultural disintegration is also quite often engendered. Furthermore, the changes in the aquatic ecosystem itself create conditions which may also threaten the health and well being of the people. On the other hand, opportunities are created which if satisfactorily managed and utilised bring much gain.

River basins have been developed to produce hydroelectric power, control floods, encourage fisheries, provide water for domestic agriculture and industrial use, establish transportation and provide for recreation.

In Africa, Lake Kariba has been created on the Zambesi, Lake Nasser on the Nile, Lake Kainji on the Niger and Volta Lake on the Volta River. These four projects by themselves have flooded over seven thousand square miles in Africa. They have each brought much good but they have also produced problems of different kinds. Increased seismic activities in the river basin, excessive deposition of sediment and silt, destruction of habitats especially for fish, and above all, the aggravation of public health problems.

The story of each project is different in its own way although some features are common to all of them. Human ecology in the context of these projects is a broad subject. The root of the problem lies deep in the basic ecological changes which occur, and they are related to altered physical, chemical and biological factors in the basin. Problems in human ecology are therefore best discussed I believe, and put in the right perspective against a background explaining the scientific basis for the changes but as I cannot do this in the available time, I shall attempt merely to offer an illustrative account and limit myself to our experience on the Volta basin development project in Ghana. Irrigation systems associated with river basins bring human problems similar to those now described.

The Volta was dammed primarily to produce hydroelectric power but it was also essentially a multipurpose project intended to improve fisheries, agriculture, water supply, transportation and even tourism. The idea of building a dam goes back some way into the history of Ghana and to around 1915. In 1953 a preparatory committee was set up under Sir Robert Jackson and the intended project was thoroughly studied before the construction started in the early 1960s. The dam was closed in April 1964.

The Volta lake became and still is, the largest single man-made lake in the world. It covers an area of about 3,275 square miles which is about 3.6% of the total surface of Ghana. It is 250 miles long on a north to south axis, stretches for about 20 miles in the widest part, and it has a shore line of over 4,000 miles. It has a drainage basin of 51,000 square miles within Ghana and this is about equivalent of two-thirds of the entire area of the country. It receives water from almost all of the country's extensive inland water system.

As a result of the dam, the immediate area which was affected by the flooding displaced 78,000 people from over 700 towns and villages of various sizes. I would like to state at this stage that whatever judgement is passed on the Volta development project, it has to be said that the resettlement project was given conscientious thought and attention. It was organised to achieve a satisfactory result and it was therefore given a multi-disciplinary approach. All stages, right from the evacuation period to the resettlement were considered in detail and preparatory investigations conducted thoughtfully by the Volta Resettlement Office. There were many problems all the way. Even the sheer size of the population which was involved was quite overwhelming. But the people were moved together with their personal effects; chairs, beds, chests, about 140,000 pieces; domestic animals, more than 170,000 including cattle and birds. Thousands and thousands of tons of food were moved and all this was transported by all means available - by land, water and by air, to the new homes.

There were also many difficulties. There were the stubborn ones who refused to believe that they could be flooded out. They wanted to stay on; they had to be taken out at the last minute. There were emotional problems as well. There were those who could not bear to leave their gods and their shrines and the graves of their ancestors and make a break with their ancestral roots. But the move was accomplished.

Before the dam closed in 1964 land had been selected and settlements constructed. Ten of these with over 3,000 houses were ready by the end of 1963 before the dam was closed. By the end of 1964, there were 11,000 houses in 44 settlements. Eventually 52 settlement towns were built and into these 69,149 people from 12,789 households were moved. Only one of the original 700 villages had had a population of over 4,000. 600 of them had less than 100 people in each village. Yet this large number of people from different places, villages and towns, different backgrounds, customs and cultures had to be put together into only 52 settlements. It was a gigantic task and whatever its failures, the organisation and the organisers must be commended for the approach which they took throughout the resettlement project.

For the selection of sites for settlements, attention was given to the accessibility of areas, suitability of the land as well as the availability of water for agriculture. Accommodation was provided in the form of core houses, with concrete floors and aluminium roofs and two bedrooms, cooking and sitting space. The core houses were so arranged that they could be added on to and extended by the settlers. The houses were undoubtedly of a superior quality and they cost up to about £330 per house but they were different from the ones which had been home for the people.

The resettlement programme was faced with a number of problems. The displaced people belonged to many ethnic groups. They had different traditions and they had different languages. They differed in education and in religion. Most of them were largely illiterate farmers and fishermen. There was also the problem of the payment of compensation to the displaced people for loss of property. Planners had intended to acquire bulk land for agriculture but the problem of evaluation and the acquisition of land became an arduous task, particularly because of the lack of relevant information and also because of poor communication. Compensation for lost assets like schools, government buildings and roads, which were vested in the Central Government presented little problem, but the assessment of compensation due within the private sector, as well as property identification and proof of ownership were quite formidable problems. Compensation, which was originally estimated at about £2 million, was paid for various items, including dwelling houses, farm land, cocoa, timber, fruit trees, fruit crops and mineral deposits.

It had been planned that the settlers were to find economic stability mainly through agriculture on family farming plots. The agricultural programme unfortunately did not proceed as satisfactorily as had been expected because of a hold-up in the land clearing scheme, and in the early stages the World Food Programme had to provide additional food for the settlers. The failure of the clearance scheme programme caused a reduction in the acreage of land for farming. In some cases, cleared areas were not ready for farming at the time the people moved into the new towns, although some experimental farms were started successfully. As part of the farming programme, the settlers were to be provided with good soil, improved seeds, fertilizers, manure, insecticides, fungicides, and even mechanisation, and the use of these facilities was to be organised on a co-operative basis. Had the clearing scheme been satisfactory, the resettlement story would have been different. By 1969, however, some of these villages were farming satisfactorily even though they had not reached the standard which had been expected.

The social problems were great for this great exodus not only meant a break with ancestral homes, it also constituted a threat to the continued stability of tradition and culture. An even more difficult problem was how to forge a socially cohesive and integrated community structure and a viable administrative system which would undertake the welfare of the new towns.

There was also the adaptation of the people and their way of life to the new ecosystem. The waters which had submerged their homes had also covered their farms and, instead of the narrow river which they know well and could confidently fish for a few thousand tons of fish annually, there was now a huge expanse of lake. The vegetation which had been submerged decomposed and provided food for the herbivorous fish. The water which was brought in by the floods contained a lot of nutrient salts which also helped with the production of plankton and algae and other plants and animals. The decomposition which initially had adversely affected the oxygen now provided abundance of organic and inorganic matter and, with the expanded area, there was a fish population explosion. By 1968, it was estimated that the lake was producing, at a conservative estimate, an annual catch of 60,000 metric tons. The African fish Tilapia, which is very popular in Ghana, became abundant. Other fishes like the butterfish, which had been hardly noticed in the river, also flourished and there were all the indications that fishing was becoming an attractive occupation. With such increased fish production, there was a potential for the lake to contribute much needed protein which would improve nutrition not only for the people near the lake and within the basin but also for the rest of the country. A resultant effect of the complex pressures which faced the people who were moved was that quite a good number of them chose to move down to the lake shore and live in fishing villages right on the edge of the lake where they could undertake fishing. Some fishermen from the coastal area of the country and others from neighbouring countries converged on the lake. By 1969, there were 20,000 inhabitants along the lake shore. It was estimated at the time that they had over 12,000 canoes. The increased size of the lake soon made their canoes rather unsafe. Their fishing methods and fishing gear were no longer efficient and adjustments had to be made to use new types of nets to fish the lake. The traditional fish preservation and processing methods which were being used, salt drying and smoking, were also investigated and improved upon to get the fishing going.

Right now fishing on Volta is a very lucrative job. The people who are actually doing the fishing are probably finding it difficult because they still have to travel long distances to catch the fish and bring the fish to the marketing place, but the amount of fish being landed is quite large.

Because of increased human activity around the lake and the ecological changes health problems also became aggravated.

Before the river was dammed, it was a breeding place of the blackfly Simulium which is the transmitter of the parasitic worm which causes the dreaded river blindness disease in Africa. The fast flowing areas of the Volta River had a number of rapids, which provided just the right type of habitat for the larval stages of the fly. When the dam was built, the breeding places were covered up as the lake rose and, above the dam the breeding places for the fly were eliminated from the river. At the same time, because the waterflow rate slowed down and there were quiet shallow shores to the lake, abundant vegetation developed.

A number of invertebrates became established in association with the aquatic plants. Among them was the snail, which is responsible for the transmission of the parasitic worm Schistosoma haematobium to man. With the frequent contact of people with the lake, and contamination of the water with the urine of people infected with schistosoma, the disease became established in foci along the shores of the lake and people have been infected. It is a serious problem, with the size of Volta and a 4,000 mile shore line, and we need a miracle to eradicate the disease from the lake shores.

On the brighter side, there has been much economic gain from the Volta project. Production of hydroelectric power was the original intention for constructing the dam. There is enough electricity for local consumption and for exporting to neighbouring countries and fish provide a high protein supply. There is abundant water for domestic and industrial use and for irrigation. There is an extensive drawdown area along the shore with fertile soil which in season can be used for short-term crops. The water level drop is about 8 to 10 feet which for a lake with a 4,000 mile shore line provides a good potential for cropping. Lake transportation is now established and there are regular services between the north and the south.

The story of Volta is a mixture of the good and the

bad. It is also, in parts, the story of Kariba, Kainji, Nasser, Kossou and the story of many other artificial lakes, large and small, existing or planned which will be in Africa and other tropical countries. For, when we undertake such major river basin development projects, and when we cause such serious changes in ecological systems, we are bound to have extensive ecological upheavals. But I have no doubt whatever in my mind that river basin development projects of this sort, whatever the disadvantages, for some countries, are necessary and desirable especially where they constitute a mainstay for economic growth. I stress this because I believe it. I also believe in an urgent need to study conscientiously and extensively to understand clearly the impact of such projects.

The problems which they cause in human ecology are complex and closely related to the basic ecological changes which take place. These problems as they affect people also deserve multi-disciplinary study if they are to be understood and solutions attempted, as indeed they should be. I have said this often and I mean it sincerely, that I am sufficiently optimistic to have unqualified hope and abundant confidence in the scientific and technical world to believe that with a bit more concern, care and deliberate will to draw on the experience which has been gained on the construction of dams we can adjust projects and minimise the ecological changes and adverse impacts caused by dams. It is not an easy task and I do not expect all people to agree with me or to be hopeful. But this is my conviction and I believe in it.

I choose deliberately to be optimistic and to transmit and convey my optimism when I can, in the hope that it will infect other people and help to forge co-operative efforts among those who have the technical ability to recognise and accept it as a responsibility and attempt to make necessary improvements, however small. At this point in time, I think it would be short-sighted of us to shirk this responsibility.

I end as I began. River basins are valuable resources; their development is vital, but they bring adverse effects. I believe these effects can be minimised, but we need a very large slice of optimism, a co-operative sense of responsibility and a judicious application of technology. We must have the social inter-dependence of people not only in countries within the Commonwealth but throughout the whole world. And we must attend to human ecology not only as it affects dam projects, but also towards creating and achieving viable political, social and economic systems. Our rivers and their basins are valuable legacies which are not mere privileges, but assets which also put on us very serious responsibilities. Our intellectual and technological resources like these assets are also capable of serving us, if we choose them wisely and with concern. We have the tools to eliminate or control much of the misery which accompanies these projects if only we find the will to do so. Dare we not take the chance? Dare we fail? It is my unqualified hope that we should not fail.

HUMAN SETTLEMENTS AND URBAN EXPANSION IN CANADA

HON B J DANSON

I All of us here today are aware that there is a broader dimension to our relationship with the Commonwealth than that occasioned by personal ties. Do not think that there is any doubt that the Commonwealth has been a success in many areas but it has been an outstanding success in the area of encouraging productive and meaningful relationships between countries of widely divergent societies and cultures. Even the severe strains which were apparent in Singapore were resolved largely because of our joint determination to ensure that the forces that bind us together prevailed, rather than the genuine differences in views which inevitably arise from our different perspectives on some matters. The Commonwealth has proved that effective relationships founded on very broadly defined goals can be lasting and, in this respect, it is a model for others to follow.

The world is in need of models of this kind. We are now in the midst of one of the most challenging and difficult periods in the history of mankind. The decisions we make in the next few years may be crucial to our survival as a world community and as individual nations.

II If one were to list the "critical issues of mankind" for the balance of this century, several would come immediately to mind: the population explosion; the food problem; energy supply and distribution; resource limits; environmental pollution. All of these issues, you will recall, have been the subject of special and urgent meetings over the past few years - meetings within nations; meetings at the regional level in Europe, the Americas, Asia and Africa, and meetings at the global level through the United Nations. We had the Stockholm Conference on the Human

Environment in 1972; the Bucharest Conference on Population in 1974; the recent Food Conference in Rome; and then, the special conferences on Energy and Resources.

There is one critical issue that provides a link between all of these and a vital (if partial) key to their resolution. That issue is Human Settlements: the shape and form and quality of our human settlements, but, most especially, the accelerating rate of urbanization and the concentration of population into a small number of very large metropolitan and megalopolitan regions.

This issue, which is in so many ways a hinge for all of the others, is to be dealt with at a special United Nations Conference on Human Settlements in June 1976. We in Canada are privileged to be hosting "Habitat", as it is called, in the city of Vancouver.

Preparations for Habitat are well underway within the UN under the guidance of a preparatory committee of 56 nations. National preparations have also been launched in dozens of countries around the world. During the past week I have been discussing Habitat and many questions that relate to Habitat with my counterparts in the Netherlands, Sweden and here in the United Kingdom. As the minister responsible in the host country, I am greatly heartened by the work that is underway. Mr Crosland, your Minister of the Environment, and I, have signed a bilateral agreement under which we will continue to exchange views on our preparations for Habitat and on a whole range of related issues of mutual concern such as urban and regional policy; housing; transportation and new communities.

At Habitat, the nations of the world will consider and, hopefully, adopt and undertake a wide range of needed international and national actions on the Global Problems of Human Settlements. This is urgently required. The changes occurring in our cities and towns and throughout our rural areas are happening at such a rate that we often don't perceive their aggregate effect. Within only twenty-five years, the terms of reference and, in many ways, the character and prospects of the human community will have changed fundamentally. Man will be living for the first time on a predominantly urban planet. If our demographic projections prove correct, three and a half billion of the world's citizens - out of a likely six and a half billion - will be in settlements of more than 20,000 people by the turn of the century. These settlements will be growing at twice the overall rate of population growth. The cities of over two to three million may well be growing twice as fast again. The dimensions of such growth are astounding. It means building as much man-made environment in 25 years as we have in the entire history of man.

Rich, developed countries in North America and Europe may be able to cope with a doubling of their urban environment in 25 years. We in fact, probably have the resources - financial and technical - to transform this rapid change into an unparalleled opportunity - an opportunity to create communities that are more conserving of energy and other resources, that are more harmonious with the natural environment, that are more human in scale and thus more livable. If we are to transform this from a crisis to an opportunity, however, we must institute measures that will enable us to manage this growth and change. Canada and other developed countries will need to develop and apply a whole range of new approaches in urban policy and institutions as much as in urban technology.

III But what of the developing world? The vast majority of mankind's new settlements will not be in the rich, settled societies. They will be in poorer, still developing lands where the resources necessary to deal with growth are tragically inadequate. When their urbanization trends are seen in the context of their population problem, their poverty, their food and energy problems, they assume the proportions of an exploding crisis. In the cities of the developing world the old environmental evils of poor water, absence of sewage and spreading slums, are coupled with modern evils of smog and fumes and chemical pollution. These cities, spreading and deteriorating over another two decades, offer us the tragic prospect of providing the very worst environment in which human beings have ever been reared.

This is a crisis from which we in the rich, developed world cannot escape. Our Prime Minister, Mr Trudeau, referred to this recently (on 13 March, 1975) in a speech at Mansion House. Referring to the old protective barriers between nations he said: "Today those barriers are gone.

There are no bulwarks behind which we can retreat in order to stave off or avoid calamity from abroad. And if there are any who believe otherwise, they are fools. Nations which are told that they can exist and flourish independent of the world are being misinformed. We are on this earth. Each has the power to injure all others. Each of us must assume the responsibility that that implies." In this crisis, the developed world has no choice but to respond. The question is not whether, but how and how quickly. Given these trends, and the needs they imply, it should not be surprising that the United Nations - at two meetings of the Governing Council on the Environment and at three meetings of the General Assembly since Stockholm, has stated that the environment problem of greatest concern to most of the nations and most of the peoples of the world is the environment of their cities and towns and villages - of their dwellings and work places - in other words, man's own habitat.

IV Work that was done for Stockholm - and since Stockholm - has demonstrated clearly that the key to the problem of resource conservation - and also the key to the problem of overloading the natural environment with waste - will have to be found largely in the better design and wiser management of our human settlements. Monitoring our atmosphere and oceans, our fish and wildlife, will tell us how rapidly we are degrading our environment. In order to stop degrading our environment, however - as you have demonstrated so dramatically here in London and with the River Thames - we must attack the sources of the wastes. Increasingly these are to be found in the economic activity and life style of our settlements. The same is true of energy consumption, resource conservation and even food supply. Let us use energy and resources, as an example. We in Canada - and most countries of the developed world - have designed our cities and towns on the assumption that the energy and other resources needed to sustain them are and will remain unlimited in quantity and cheap in price. Look at our recent urban systems; isolated rather than community heating systems: high rise towers; sealed, air conditioned, with complex vertical transportation systems: our urban systems are highly consumptive of energy and other resources and they generate an increasing volume of waste that imposes an intolerable burden on the land and on our common atmosphere and oceans.

We don't need to continue to build such systems. With

present technology, we can design urban systems that are far less wasteful of energy and resources without reducing either our standard of living or the amenities that we enjoy. We can also significantly reduce the social problems and the alienation that is increasingly characteristic of our cities. In short, we can find a new synthesis between man's continuing desire for betterment and the constraints of a finite world with finite resources.

This will require a many-faceted approach. It will require new policy and institutional approaches to the management of urbanization and to the planning of future communities. More importantly, it will require that we identify and apply the best of the available approaches around the world, that have been found successful and that are more or less transferable by and to other countries. We need to know more about the successes, and the failures, of one another so that we can better deal with our own problems. Habitat, and the preparations for Habitat, are intended to provide this opportunity.

V Western Europe - and especially you in the United Kingdom - have pioneered many of the more successful approaches to urbanization. Your regional planning and development policies, your new towns, your public transportation are examples of this. I have been discussing these and other approaches with European Ministers and officials and I have seen examples of where and how they have been applied. I have been impressed both by the similarity of our problems and by the relevance to Canada of so many of the approaches that have been tried.

This is of immediate interest to us in Canada. We are in the process of re-examining Canadian urban trends and the types of policies and programmes needed to shape these trends.

Our urbanization trends are not too different from those of Western Europe, the United States and other developed countries. In brief - we face a period of extremely rapid urbanization - perhaps a doubling of our total urban environment in less than 25 years. We are disturbed by this projected pace of urbanization because we feel that it could overwhelm the capacity of our institutions to plan for it and absorb it in a manner that would produce a livable and high quality urban environment. We are perhaps even more disturbed by the projected distribution of this urban growth. If the trends are allowed to unfold, Canada faces a future that is not only predominantly urban but also one in which the overwhelming majority of Canadians will be concentrated in a small number of very large urban regions. Indeed, by the turn of the century, two-thirds of our total population would be living in just three provinces, Ontario, Alberta and British Columbia, and three-quarters of that two-thirds - or roughly half of all Canadians - would be living in the Montreal, Toronto and Vancouver-centred regions.

If allowed to unfold, these trends would have an immense impact not only on these provinces, and cities could become unmanageable; the others, continuing to lose population, would wither. Regional economic disparities would be accentuated and political power would shift, perhaps to quasi City-States, but most certainly to the dominant provinces, to an even greater extent than today.

The government of Canada, the governments of all our provinces and most of our cities have agreed that we must co-operate in the development and application of policies to shift these trends toward more desirable objectives. Just before leaving Canada, I launched a process of intergovernmental and public consultation on a national urban strategy. Together we are looking at the objectives we want to pursue and at the types of policies and programmes needed to achieve these objectives.

It is very important, even though politicians must ultimately make the decisions, to expose the thinking to the widest possible range of people, so that when we achieve a concensus, if that is possible, then the basis of those decisions will be understood, whether agreed with or not.

Basically, and briefly, we are looking at a strategy that embraces objectives and policies in three inter-related areas: first, the future size of Canada's population and its rate of growth. Second, the distribution of our future population across Canada and in our urban communities. Third, the management of our future urban growth so as to create the kind of cities and communities that we want. These are difficult policy areas and effective responses in them will not be easy to provide. But they will be provided, implicitly if not explicitly. Given our federal system, it is desirable, if not necessary, that they be provided explicitly.

In Canada we are asking ourselves: if the patterns of growth being unfolded by the trends are unacceptable, what alternative patterns of growth would be desirable? What public spheres can best be used to achieve them? And what public policies can best be used to create communities that are livable, human in scale and in harmony with the natural environment? The responses to these questions cannot be imposed by any level of government. It is essential to achieve a broad national consensus on the objectives we are to pursue. Once we have that consensus, we will need to determine the best means available compatible with our value system and our democratic form of government.

Canada is not unique in asking these questions nor in searching for appropriate responses to them. That is evident from the five tentative themes for Habitat adopted by the United Nations Preparatory Committee in January. The first of these, in fact, is "National Settlement Policies and Development". The second is "The Social and Economic Aspects of Settlements". The third is "The Planning and Management of Settlements". The fourth is "The Design and Construction of Shelter and Services". And the fifth is "Human Settlements and the Natural Environment".

Virtually every member country of the United Nations faces the problem of rapid urbanization and the need to manage urban growth. When it comes to means, we have a great deal to learn from one another. We in Canada are especially interested in the experience of Western Europe and the Commonwealth, with many of whom we share value systems and forms of government.

VI Each country, of course, has to develop its own response to urbanization and its own means to manage urban growth. The scope for international action on the problems of human settlements is limited. The really vital actions needed to solve these problems must be undertaken by countries themselves. This is as true for developing countries as it is for developed countries.

That is why, in the preparations for Habitat, nations have agreed to spend a great deal of time and effort in identifying approaches to human settlements problems that have been applied in one country or region and that may have elements that are transferable to other countries or regions.

I believe that through this kind of exchange, people and nations and governments will see that human settlements problems are capable of solution - that solutions are indeed available if we have the common will and wit to apply them - that we don't need to shirk from addressing these problems.

Although the scope for international action on human settlements is limited, that which can be taken is vitally important. Between now and June 1976 countries will be working separately and together on the development of recommendations for international action. The Government of Canada has no firm view on the question yet and, of course, will not adopt one until the end of the preparatory process when we have had the benefit of advice from other governments, non-governmental bodies and private citizens. But we do have some preliminary views that I would like to expose.

VII In my view, at this stage, Habitat should make a significant advance in at least four areas.

First, Habitat should have a number of important programme results. These could include, perhaps, a decision to have an on-going UN human settlements demonstration programme. If a concrete programme for the exchange of information and ideas on human settlements were established, it would be of tremendous benefit not only to those nations where urban problems are of the greatest concern, but also to Canada, the United Kingdom, and other developed nations.

Another area where Habitat should have important results is in education and research. These, perhaps, could include a decision to strengthen and establish a number of regional urban management training institutes. There is an acknowledged need to better the competence of urban management throughout the world both in the developed and developing nations. And I think it is essential to the future of human settlements that nations develop and train leaders and officials who can grapple with the task of managing the huge cities that are an inevitable part of our future.

Another result was called for when the General Assembly

launched Habitat. The Assembly requested that the conference should have a "financial" and an "institutional" result. Canada recognizes that this is very important. We also recognize that any recommendations in this area, to be meaningful, will require the most careful consideration and must carry the broadest possible measure of support from governments.

Finally, I would like to see Habitat adopt a firm declaration of principles with three basic characteristics. It should recognize the fact that human settlement is one of the critical issues of mankind. Secondly, it must recognize the diversity and complexity of human settlements and it should identify the main areas of action as well as the political and scientific resources that need to be marshalled. Thirdly, it should represent a commitment by governments to tackle human settlements issues with the resources and urgency that are required.

I realize that this is a tall order. But this is what Habitat is all about, and in developing and refining such a declaration of principles, it seems inevitable that our understanding of human settlements issues, and the commitment by our governments to their resolution, will be strengthened. And this will benefit all nations.

The challenge is immense, but it will not disappear. Indeed it will intensify and demands the most thoughtful, cohesive and energetic applications of our diverse and disparate resources. Habitat presents a unique and timely opportunity to harness the genius of man to meet this challenge.

HUMAN ECOLOGY AND DEVELOPMENT: CANADA AND THE NORTH AMERICAN REGION

LEONARD O GERTLER

From what I know about the impact of the Honourable Mr Danson's lecture, his presentation is a hard act to follow. I hope, however, that I will be able to reward your gracious tolerance of two Canadians speaking in succession, by attempting to draw out the implications of one of his central themes, namely Canada's search for a national urban strategy.

While the Minister with his legendary persuasiveness has made the initiative appear as natural and inevitable and irresistible as Niagara Falls, it is as an exercise in policy-making an event with a deep and contentious history, and most likely, a no less fractious future. We have in fact travelled a long hard road; and I will attempt to interpret for you some of the landscape along the way, as well as some of the besetting hazards - slippery pavements and assorted highwaymen - that have marked our advance towards the Minister's declared destination: "a livable and high quality environment". The motives for this backward look are, of course, not antiquarian. My concern is to throw light on two related questions:

How have we come to this point of illumination and apparent decisiveness?

And what are the conditions and prospects for success?

In this exposition I shall have recourse to three concepts which, in the cause of communication, I would like to define - very briefly. These are 'growth', 'environment-making' and 'development'. <u>Growth</u> refers to expansion - of production, human settlement, or an entire system without any basic change in its structure. <u>Environment-making</u> is the process of creating a setting in which both the biotic and built components are favourable to the long-term well-being of human communities. And <u>development</u> refers to "the unfolding of the creative possibilities inherent in society", and it usually requires 'innovation' - changes which are widely perceived as fundamental (1). In the lexicon that I will use 'development' occurs only when 'growth' is accompanied by 'environment making'.

The subject of the "human consequences of urbanisation" has very recently had the benefit of a diagnosis by the noted American Geographer, Brian Berry. Canada, according to that interpretation, falls into the same tradition of free enterprise growth dynamics as the United States, and as a society which tolerates a rich pluralism is inherently 'incapable of being goal-orientated' (2). While there are now rumblings of change under the banner of post-industrialism, it is a tradition which exalts "privatism" - the priority of private individual and corporate decision-making as well as consumer gratification, inhibits community action, and which only reluctantly concedes the need in times of crises for positive public intervention. The bedrock functions of government action remain "to protect and support the central institution of the market and to maintain the required dispersion of power" (3).

This vision of reality is contrasted with other major models that move along a path towards social purpose and social action; for example, the redistributive welfare states of Western Europe and the Third World countries both of which are tending, for different reasons, towards the formulation of broad, goal-oriented national urban policies. In action form, these are characteristically expressed as strategies to shape the regional distribution of industry and population, while protecting fragile environments and aspects of traditional cultures, and steering future growth away from areas experiencing environmental overload (4).

While Canadians have often, for both legitimate and other reasons, been assigned to the same bed as Americans, I believe in this case that the attribution is not quite deserved. There is a difference and on that difference hangs much of our chances for moving from the pursuit of growth for its own sake to a policy of national development.

Consider for a moment the differences between the American and Canadian nineteenth century railway building eras. In the first, you had a frenzy of competitive rail construction by numerous companies whose colourful names are still celebrated in American folk song. In the second you had a few regional and national undertakings, predestined to become the C.P. and C.N. In America, you had the Robber Barons - and the tooth and claw battles for control such as the War of Erie between Commodore Vanderbilt and Jay Gould (5). In Canada, the images that survive are Lord Strathcona and the last spike, and Beatrice Webb's papa, Richard Potter, riding in elegant and genteel splendour in the president's car on a tour of inspection of the Grand Trunk Railway (6).

The creation of the transcontinental railway was an act of state. This was not the first but the most conspicuous early expression - in fact, it was one of the terms of union with British Columbia, of the willingness of Canadians to have their government undertake in the national interest, what the enterprise system on its own could or would not do. And this - the deployment of public power for transcending public purposes has been one of the continuing motifs of the Canadian story.

But it has been only one of the motifs. Canadian history looked at through the spectacles of the human ecologist has in fact gyrated between economic privatism and social responsibility. And the economic tradition is still very much alive. Just a few weeks ago McClelland and Stewart published a book by the Carleton University sociologist, Wallace Clement, which argues, with some conviction, that Canada enjoys the mixed blessings of an indigenous economic elite that has its origins in successive foreign investment in the country's staple products and natural resources by metropolitan interests in France, England and U.S.A. (7). An original go-between or entrepot role has been parlayed into a dominant position in the strategic sectors of finance, commerce, transportation and utilities. As a consequence, the ideology of privatism is identified with substantial corporate influence.

Without, however, anticipating the denouement of my plot too much I would like to illustrate by reference to a few highlights the continuous interplay of growth, environment-making and development in the Canadian flow of events.

At the time Theodore Roosevelt raised the banner of Conservation, Canada reacted to its first fifty years of resource exploitation by establishing a Commission of Conservation, which for a dozen years, 1909 to 1921, pursued its broad mandate of research and public education on the use and protection of natural and human resources.

The Commission's Advisor on Town Planning was a practitioner from England, Thomas Adams, a contemporary and friend of Patrick Geddes and Ebenezer Howard, who over a period of seven years made a far reaching impression on the Canadian scene. Working through the Commission's Committee on Public Health he laid the foundation for planning legislation in seven provinces, and in his book on Rural Planning and Development (1917) developed a concept of planning, which encompassed both town and country, both the well-being of people and the management of natural resources and which after some years of eclipse has become a living tradition in Canadian concepts of regional planning (8).

The Commission as a research, conference, educational vehicle became the model for a series of national appraisals of man, resources and environment in 1961, 1966 and 1973, which together marked the country's evolving concerns - from the waste and mismanagement of resources, to economic development and planning, to the quality of environment.

However, even while the ink was wet on the Commission's pronouncements a major miscalculation in land-use was underway in south-east Alberta where over twelve million acres of grassland were plowed up and settled in an area which Captain Palliser, a Colonial Office surveyor, had characterized as a 'semi-arid desert' (9).

The region was occupied during a freak wet phase of the hydrological cycle in the general euphoria of pioneer settlement, propelled by the powerful attractions of cheap land and speculative profits. As Chester Martin has explained: "With every sign of improvement in general conditions .. the land speculator was there in advance of the farmer, the village doctor, lawyer and store-keeper. Only too frequently he was the village lawyer or doctor or store-keeper himself" (10).

Almost as quickly as the whole apparatus of a western Prairie town/country community was created it was swept away by the cruel dry winds of the plains which refused to yield up the moisture necessary to sustain a grain regime, leaving behind after the seven fat years only ghost towns and the bitter memories of what came to be called "next year country" - next year things would be better.

This was followed by the enactment of one of those endemic Canadian melodramas in which the community closes ranks after a binge of privatist excess, and various forms of co-operative action are invoked to stave off disaster. Most of the stricken settlers of the Palliser triangle were given government assistance to resettle in more secure park-belt black soil areas to the north - and lived happily ever after.

This dualism - of individualism and co-operation - of the pursuit of private interests whatever the consequences counterpointed, under suitable provocations, by the remorse of social responsibility has entered deeply into the Canadian experience - and maybe soul, if we have one.

It was expressed in the first two decades of this century when the Prairie grain farmers - rugged individualists par excellence - formed co-operative grain companies to protect themselves against manipulation of storage rates and grain prices.

The initiator of this movement, E. A. Partridge, was a true representative of his times: he moved from pioneer sod-hut settler to co-operative leader to advocate, in a publication of 1925, of a rural utopia in the form of a network of self-sufficient and self-governing communes of between thirty-five hundred and seven thousand residents, without private property, rent, taxes or lawyers (11).

The Canadian novelist and critic, Henry Kreisel, throws further light on the dualism - not quite schizophrenia - of the Canadian social psyche. "The conquest of territory", he wrote, "is by definition a violent process. In the Canadian west, as elsewhere on this continent, it involved the displacement of the indigenous population by often scandalous means, and then the taming of the land itself". And he goes on to illustrate that the price paid is an ego-centric, all-consuming obsession with the possession of land - "perhaps necessary", he says, "if the huge task of taming a continent is to be successfully accomplished. At the same time, the necessity of survival dictates co-operative undertakings. So it is not surprising that the prairie has produced the most right-wing as well as the most left-wing provincial governments in Canada. But whether conservative or radical, these governments have always been puritan in outlook, a true reflection of their constituencies" (12).

Now I do not wish to draw out this bow of historical interpretation too far or in too much detail.

Suffice it to say that the congenital melodrama was reenacted during the Depression and reinforced by the years of World War II; and has by now entered deeply into modes of thought as well as political institutions and concepts. Its more recent manifestations have dramatized the structural faults in the Canadian system - and have focussed on the issue of regional disparity and maldistribution: economic, social and environmental.

A line of thought, policy and action can be drawn that links the mid-thirty efforts to shore up the economies of the Prairies and the Maritimes, including some quite massive resource-oriented programmes of irrigation, soil conservation and marshland reclamation; the welding into our fiscal structure of a system of equalization grants - the Robin Hood principle of taking from the rich and giving to the poor; the Economic Council's identification of balanced regional development as a goal of national policy; the priority given in the 'sixties to programmes of rural adjustment and rehabilitation by both provincial and federal governments: Quebec's quiet revolution; the massive urbanization of the past quarter century; the rise of an environmental conscience; and the emergence at the national level of three broad coordinative policy fields: Regional Development, Environment and Urban Affairs (13).

I do not wish for a moment to convey the impression that these happenings and events have in any sense been linked together by a smooth and rational process.

It was not so long ago, for example, that a Governor of the Bank of Canada was offered a premature pension - an offer he could not refuse - for wishing to address the powers of the monetary system to the overcoming of regional disparity.

But I do wish to say that when a Canadian Minister declares - as he has recently on this platform - that the government is giving the highest priority to the working out of a national urban strategy - it is a symptom of some fairly deep stirrings in the Canadian body politic - as well as an unparalleled opportunity to coax the Canadian system away from growth to environment-making and development.

I shall in a moment, turn to some conjecture on our chances of making it to the new Jerusalem.

But first it is necessary to enlarge our perspective by bringing into focus some of the external influences working on Canada's growth and development.

Canada's external relationships have been a major force in shaping the centralized pattern of development that has so justifiably aroused the concern of the Minister of State for Urban Affairs. The Atlantic region, in its urban structure and growth pattern, may be seen as a by-product of the rise and fall of its shipping predominance - the mid-nineteenth century era of 'wood-wind-water', based on triangular trade patterns between Canada, the Atlantic seaboard and West Indies, and the United Kingdom. International trade also established the predominance of Montreal, the first big entrepot city; opened the West by linking the grasslands to world grain markets; and catapulted Vancouver into metropolitan orbit after the opening of the Panama Canal in 1915 (14).

A second major external influence, foreign investment, has worked in two directions, towards the extractive, northern hinterlands to 'harvest' the pulp and paper, base metals, iron and hydro power; and towards manufacturing, mainly in Ontario, to serve the expanding national market and, through subsidiaries, to reap the advantages of Canada's preferential trade arrangements.

Immigration, a third major aspect of Canada's global relationships, has assumed the characteristic of several major waves, beginning with the Loyalist exodus during the American Revolutionary War and then occurring in six major periods to the present; and following certain characteristic patterns, in terms of origins and settlement. Generally Britain and Europe have been major sources, but in recent years there has been a striking increase in immigration from the United States, Asia and the Caribbean. In 1973, when 184,000 arrived, included in ten leading source countries of immigrants were Hong Kong, Jamaica, India and Trinidad. Compared to the 1950-55 period, immigration from Europe dropped from 88 per cent to less than 50 per cent in the five-year period 1968-73; while people from Asia increased in the same period from 2.8 per cent of the total to 16.8 per cent and from the Caribbean and Central America, from .7 per cent to 8.4 per cent (15). As for the destination of immigrants the tendency of settlement has been over the years from the rural areas during the period of large-scale pioneer settlement to the bigger towns and cities, with a really dramatic concentration in the last period in Toronto, Montreal and Vancouver which attracted in the years 1966-71, about 41 per cent, 16 per cent and 11 per cent of the total, respectively (16).

I have referred to these broad relationships - in trade, investment and immigration - because each of these represent certain trends and forces that impact on growth patterns and which constrain our capacity to carry out national policies that confront the chronic concentration of people and economic activities in a few big places. This particular policy preoccupation involves us deeply in the relationship between growth, environment-making and development. In Canada, urban incomes are significantly higher than rural incomes. This is illustrated by certain inter-regional comparisons. For example, the 1971-72 earned income per capita, as a percentage of the Canadian average, is 66 per cent for the Atlantic provinces and 119 per cent for Ontario - a difference of 53 per cent; but the difference in average income for the cities of the two regions is only about 18 per cent (105.2/87.3 per cent) (17).

It should not be inferred, however, that increases in per capita income rise in proportion to increases in city size. In the Canadian system, data suggests that there is a tapering off around a population level of 200,000; at the same time this also appears to be the point where municipal expenditures rise sharply. Underlying this are certain environmental and social stresses, which show up particularly in such indicators as air pollution (sulphur dioxide and suspended particulate matter), and the rate of legal offenses which over a considerable period of years has been consistently highest in cities beyond a quarter of a million (e.g. 1970 : 250,000 and over 13 per cent greater than municipalities 100-250,000, and these 13 per cent greater than the next size group 50,000-100,000 and so on) and has been strongly correlated with city size. But beyond the mystification of statistics there are deeper issues related to the sharing of opportunities for the 'good life' and the political complexion of a country that is a sub-continent which could have, if we let things ride, over half of its population concentrated at three points (18).

Returning now to the reverberation of external influences there are certain aspects of foreign investment and immigration which are particularly important for this discussion. During the past quarter century American investment has become dominant in Resources and Manufacturing - for example, about 69 per cent of Mining assets are foreign controlled and 80 per cent of this is U.S., and about one-third of dominant directorships in manufacturing are American compared to 45 per cent Canadian and 12 per cent U.K. (19).

Resources investments must follow dispersed resource patterns, but U.S. controlled manufacturing establishments gravitate to one region: south-western Ontario, reluctant apparently to extend the umbilical cord too far from New York and Chicago head offices, and lured by the high market potential of the Toronto region - 83 per cent of U.S. controlled manufacturing employment is within 400 miles of Toronto. This tendency is not as strong for other foreign controlled employment - for example, U.K. firms control a greater part of the employment of the Atlantic provinces than U.S. firms (15 per cent compared to 6 per cent) and dominate only 6 per cent of Ontario employment compared to the American figure of 31 per cent (20).

Overall the effect of foreign investment in Canada appears to be to accentuate the basic concentration of the Canadian economic structure, documented a decade ago by John Porter and further elucidated this year (in 1975) by Wallace Clement. While a facile association between economic concentration and geographical centralization cannot be made - the link just has not been established - there is 'circumstantial evidence' which at the very least provides material for a respectable hypothesis. Clement, for example, presents from official data the following: the five largest corporations in their respective fields control 93 per cent of the assets of chartered banks, 63 per cent of Life Insurance Companies and 61 per cent of Trust Companies. And he postulated that the financial sector is at the apex of an interlocked network that includes manufacturing, resources, utilities and trades. He further indicates that in these sectors there are 113 dominant corporations managed by a group of 946 Canadian residents and 287 foreign residents and that these individuals form a self-conscious, self-perpetuating elite knitted together by ties of kinship, club, school, church and philanthropy. A most interesting point about the clubs is that there are only six and that they are all in central Canada - three in Toronto, two in Montreal and one in Ottawa, and that over half of the economic elite belong to one or more of them (21). So, here we have emerging the picture of an economic elite in which Finance - the lubricant of the whole system - has its skyscraper temples concentrated in Toronto and Montreal, where they can find solace and reinforcement on certain certified social ground described as meeting places "where businessmen can entertain and make deals", and as "exclusive gentlemen's clubs ... the preserve of the upper class male" (22).

I can vouch for the last part of this characterization for I have had the pleasure of being invited to lunch to the National Club by a lady, a most accomplished member of the Legislature of Ontario. And to our mutual embarrassment and consternation, I was allowed in by the front door and she, a member, crept in by the serviceman's entrance. So much for the <u>mores</u> of Canada.

As far as immigration is concerned it is of some interest to note that the pronounced Ontario bias of immigrant destination has not been exclusively due to 'push factors' from countries of origin. In a refreshingly candid publication, released in 1972 by the Ontario Economic Council under the title of <u>Ontario, A Society in Transition</u>, Don Richmond, its author, makes it quite clear that a high level of immigration was sought in the 'fifties and 'sixties to fill the labour gap left by the low birth rates of the depression 'thirties and early 'forties. He states: "In Ontario, the inflow of immigrants coincided with the urbanization of our population and the industrialization of our economy ... For those who insist on a cause-effect explanation of economic events, the inter-relation between immigration, industrial development and capital inflows becomes a chicken and egg argument. A linear explanation is not adequate. What we have to deal with are three distinct but related variables all interacting and all moving in a single direction. The direction is economic growth and economic diversity"(23).

Canada at the present time is in the midst of a public debate on immigration, precipitated by a Green Paper on Immigration Policy, tabled in the House of Commons on February 3rd, 1975. The paper puts forward two propositions of far-reaching significance for Canadian development. One is that the country's fertility rate (1.9, the average number of births per woman) has attained and is likely to remain at the population replacement level, making immigration the critical factor affecting the future growth rate of Canadian population. The other is that "future developments respecting population distribution and concentration can be expected to have a more immediate impact on the well-being of Canadians over the next few decades than will aggregate national population growth rates" (24).

With this perspective, immigration policy merges into population and urban policy affecting the national pattern of human settlement. This is further reinforced from recent findings that Canadian migrants and immigrants tend to have similar preferences concerning the places where they want to live and work in Canada. For example, it is estimated that about sixty-five per cent of recent West Indian immigrants have settled in the larger cities of Ontario (25). Thus the distribution aspects of immigration are central to the manner in which we sort out the relationships between growth, environment-making and development.

Before directly addressing the distribution question there are several conditioning factors that must be briefly stated. One is the constancy of certain key elements in Canadian immigration policy which were reiterated by the Minister of Manpower and Immigration, the Honourable Robert Andras, in his explanatory statement to the House of Commons. These are non-discrimination by reason of race, colour or religion; respect for family immigration; the acceptance of refugees on compassionate grounds; and the needs of economic and social development (26).

A second factor is the commonality of human motivation underlying mobility, whether it is internal or external. One of Canada's foremost demographers, Leroy Stone - a Canadian of West Indian origin - stresses the importance of the personal urge for improvement, "the existence of a system of strategies adopted by the individual in the course of passing through the life cycle" (27). There is a bond - the search for education and better work and housing - between the men and women and families moving in Jamaica from Mandeville to Maypen to Spanish Town to Kingston - and sometimes to Montreal and Toronto; and similar groups moving in Canada from Daupin to Portage La Prairie to Winnipeg.

While we recognize this common humanity in our immigration policy, the same principle is not apparently always applied at the societal level within Canada. Ramcharan of the University of Windsor finds in a recent sample survey that 70 per cent of West Indians who have been in Canada less than a period of seven years hold a perception of being discriminated against in employment - although we may take some encouragement from the finding that this perceived discrimination drops to about forty-eight per cent of the group after a residence of more than seven years, and that about three-quarters of this group of more established residents report a general satisfaction with their lot in their adopted country (28).

These factors have been emphasized, not only for their Commonwealth interest, but because given the global population dynamics of this epoch the much more diversified origins of the immigrant flow to Canada cannot be regarded as a passing phase. The search for a less concentrated and more livable urban pattern in Canada leads to an inescapable question of increasing urgency that we have to ask the person from England, United States, France, Portugal, India and Jamaica: "Will you settle in Trois Rivieres, Thunder Bay or Saskatoon instead of Montreal, Toronto or Vancouver if you have a reasonable chance for jobs in those places, and perhaps better housing and other services (because these can be directly influenced by public action), plus the prospect of a life with less conflict and stress than the life of the big city"?

In the broad historical sense, the immigrant to Canada has always answered this kind of question in the affirmative, and has given generously to the opening up of our farm lands and mines, the building of our railways, the development of our towns and cities, as well as art, science and industry. In the present period there are some special factors operating that offer the promise of success in a policy of decentralization - the holy grail that is the pursuit of the Ministry of State for Urban Affairs. These are a) the present global climate of very strong interest of prospective migrants to come to North America, b) the greatly increased awareness, in this post-Bucharest world, of the problems of population growth and distribution, c) and the emergence in Canada at this time of a new maturity in policy thinking as well as delivery skills, as in the area of regional development, which go in the direction of identifying the strategic determinants of the country's development/settlement structure and mobilizing policy/programme resources towards a limited number of objectives having powerful leverage on the entire system (29).

It would be a pleasure to conclude my remarks on this note of high optimism. But these must be tempered by the other considerations that I have attempted to bring into focus: the conflicts that run deep in Canadian society; the stubborn persistence of a concentrated settlement pattern and regional disparities; the biases of economic forces shaped by a centralized metropolitan corporate elite. To these must be added the effects of our system of federalism, which in Canada is certainly more than a word, and which is a way of political life which has the virtue of highlighting the regional fact and preventing power from being too corrupting; but which also involves a struggle for dominance which sometimes transcends and confuses the real life issues of work and recreation and environment that really matter to everyman.

There are, in Canada, some detractors, like Professor Harvey Lithwick, the author of Urban Canada, who has played an important role in the evolution of Canadian urban research and policy, who invoke a "corporate state" image in which public and private technologies reinforce each other and are expressed in a powerful and irresistible sectoral stance - what's good for General Motors is good for Transport Canada and so on (30). But I find that I cannot identify with that Stygian view, if only because I have had the opportunity, on occasion, to leave the classroom and join the direct struggle for betterment and have gained in that edifying, but painful process, some confidence in the constructive forces in Canadian society: a reservoir of energy and assertion of humane values which are a constant source of amazement - and reassurance. But, in the final analysis, I must leave to you, and unfolding events, the adjudication of the issues that I have raised.

NOTES

(1) John Friedmann, Urbanization, Planning and National Development, Beverly Hills/London: Sage Publications, 1973, pp.43-48
(2) Brian J. L. Berry, The Human Consequences of Urbanization, London: MacMillan, 1973, p.71
(3) Ibid., p.166
(4) Ibid., pp.142 and 171
(5) Matthew Josephson, The Robber Barons, New York: Harcourt, Brace and Company, 1934, pp.121-148
(6) Beatrice Webb, My Autobiography, London: Longmans Green & Co., 1946, pp.4 and 150
(7) Wallace Clement, The Canadian Corporate Elite, Toronto: McClelland and Stewart Ltd., 1975, p.XII
(8) Thomas Adams, Rural Planning and Development, Ottawa: Commission of Conservation, 1917
(9) Jean Burnet, Next-Year Country, Toronto: University of Toronto Press, 1951
(10) Chester Martin, "Dominion Lands" Policy, Toronto: The MacMillan Company of Canada, 1938
(11) Heather Robertson, Grass Roots, Toronto: James, Lewis and Samuel, 1973
(12) Henry Kreisel, "The Prairie : A State of Mind", The Canadian Anthology, edited by Carl F Klinck and Reginald E Waters, Toronto: Gage, 1974, p.624
(13) Leonard O Gertler, Regional Planning in Canada, Montreal: Harvest House, 1972, Chs.1,3,4,7 and 8
(14) Encyclopaedia Britannica, Chicago: William Benton, 1964, P.971
(15) Immigration and Population Statistics, Ottawa: Manpower and Immigration, Information Canada, 1974, p.31
The Immigration Program, Ottawa: Manpower and Immigration, Information Canada, 1974, pp.83-84
(16) Ronald W Crowley, Distribution of Population, Ottawa: Ministry of State for Urban Affairs, 1975

(17) D Michael Ray and Paul Y Villeneuve, Population Growth and Distribution in Canada : Problems, Process and Policies, Ottawa: Canadian Council on Urban and Regional Research, 1974, pp.40 and 43

(18) Local Government Finance - Revenue Expenditure - Preliminary and Estimation, #68-203, Ottawa: Statistics Canada, 1972
Perspective Canada, A Compendium of Social Statistics, Ottawa: Statistics Canada, Information Canada, 1974, pp.197-198 and 292

(19) Op. cit., Clement, Table 13, p.144

(20) Op. cit., Ray, p.48
D Michael Ray, "The Growth and Form of Urban Centers in Southwestern Ontario", Trends, Issues and Possibilities for Urban Development in Central and Southern Ontario, Toronto: Ontario Economic Council, 1970, pp.18-20

(21) Op. cit., Clement, pp.110,160,167,247-48

(22) Ibid., p.247

(23) Don Richmond, Ontario, A Society in Transition, Toronto: Ontario Economic Council, 1972, p.76

(24) Immigration Policy Perspectives, Ottawa: Manpower and Immigration, Information Canada, 1974, pp.4 and 8

(25) Subhas Ramcharan, The Economic Organization of West Indians in Toronto, Ottawa: A Report Submitted to the Department of Manpower and Immigration, 1973, p.iii

(26) Freda Hawkins, "A Report on the Green Paper on Immigration Policy", Urban Forum, Spring, 1975, p.8

(27) Leroy O Stone, "What We Know About Migration Within Canada", Conference on Migration of the International Sociological Association, Toronto: October, 1973, p.25

(28) Op. cit., Ramcharan, Part 2, p.10; Part 1, p.24

(29) Op. cit., Hawkins, p.12

(30) N H Lithwick, "Political Innovation : A Case Study", PLAN, Canada, Vol.12, No.1, 1972, pp.53-54

ECOLOGY AND DEVELOPMENT - SOME ASPECTS OF THE JAMAICAN EXPERIENCE

HON ALLAN ISAACS

I have been asked to say something about what is termed Human Ecology - to give some account of the Jamaican and Caribbean thinking on the subject.

But from this perspective, my function is to present my own diagnosis of the present situation in terms of the ecology of this world. I am deeply convinced of the fact that ultimately there is one ecology, and that is this planet; of the fact that ecology contains numerous, numberless, other ecologies; that all ecologies are about one or other form of life; and that man, being prime creature on earth, being the most intelligent, the most ingenious, has made all the other ecologies his own. By doing this he has acquired a very grave responsibility. First, a commitment to a respect not only of human life but for life itself. Without that idea ecology is meaningless. And one crucial fact about ecology is that any diminution of the respect and regard for life is a diminution of the quality of life of our own species and the diminution of our own chances of survival.

It is necessary for clarity to say what I mean by ENVIRONMENT. The ENVIRONMENT is, for our present purposes, the place - the habitat in which MAN lives and all the features of that place; these features include all things living and inert and the effect of the interaction between all these components or factors <u>including Man</u> himself.

Ecology is in the present context the name attached to the interaction between the several components or factors of the Environment.

In HUMAN ECOLOGY, of course, Man is the prime component, and, indeed, the determinant of the character of the system.

It is a truism (which unfortunately is often overlooked) that each of the organic components of HUMAN ECOLOGY - that is to say, plants, animals, birds, fish, insects, soil, water, algae, lichens, amoebae and so forth, have distinct ecologies of their own. Each of these components survives not merely through tolerant and passive co-existence with others, but through dynamic interaction and vital interdependence between them. Distinct group interrelationships are called eco-systems, so that at any given moment Human Ecology is equal to the sum of all the ecosystems within the human environment.

The elimination or even the diminution of any one component diminishes and endangers the others. Man himself, because of the increase in his awareness and the growth of his intelligence is not merely a component of his own ENVIRONMENT but the PROPRIETOR of the others. With the growth of knowledge and the spatial expansion of his power to interfere, Man has enlarged his environment to the limits of the earth with all its contents, inert and organic alike.

Man's technological inventions have enabled him to traverse the earth and to take possession and do his will on sea and land alike at any distance from his own dwelling place. The major examples of this phenomenon are, of course, geographical exploration, discovery, conquest and settlement. These have brought fundamental changes virtually over the entire planet.

All continents were, for instance, extensions of the European imperial domains and were subjected to irreversible ecological changes and became by the same token, extensions of the HUMAN ECOLOGY of Europe to a very substantial degree. The age of territorial discovery and conquest is evidently giving way to an age of territorial liberation, but, with the advent of territorial independence, ecological change seems destined to assume new dimensions.

That is because whereas there were only a few motive forces, the British Imperial System, the French Imperial System, the Spanish Imperial System, a few systems all originating in Europe, which had the effect of extending the relevant ecologies to various parts of the world, now that the liberation movement is practically at an end, each nation has got sovereignty and responsibility over its own environment, and is free to interfere according to its own likes and its own drives.

Man has long been aware that the earth is his estate and that he has the power to exploit it for satisfaction of his own needs and desires including his mere pleasure. What is crucial and urgent today is that Man should readjust his view of this planet as something to be casually exploited, to that of a heritage to be husbanded and cared for and used for his own good. The good of all other species and the integrity of inert features are ultimately indispensable to his own survival and his own welfare.

The contents of the earth have come in recent times to be known as NATURAL RESOURCES which are divided into six broad categories, LAND, WATER, ENERGY, AIR, FLORA and FAUNA. Man, himself, falls within the category of FAUNA. Each of them incorporates many species, classes, types, kinds, varieties and strains, and their care is intricately dependent on the <u>inter</u> and <u>intra</u>-relationships between them.

These resources are diverse, but there must be harmony between them which it seems to me Man should comprehend and ultimately realise without disturbing in any significant way the prerogative of any of the species to obtain what it needs to exist and, indeed, to flourish.

Man can comprehend this proposition and has the natural endowments of mind and spirit that are required to implement it. It is the deliberate and systematic process by which the realization of this ideal is pursued that I understand as RESOURCE MANAGEMENT.

It is evident that even in a much shrunken world this process cannot be applied everywhere at the same time, or that it can realize the ideal in any given place immediately. The supremely urgent task is that the recognition and acknowledgement of this human obligation should be propagated as rapidly as human energy and human ingenuity will allow. This must be conceded if HUMAN ECOLOGY is to be a single composite ecology with an essential validity throughout the world. To myself this is clear.

The main ecological aim of husbandry as distinct from mere exploitation is to ensure that development and production do not devastate or sterilise the environment. Development will, no doubt, modify or even transform the environment since it usually introduces new features; but the acid test is that the consequence will be an ecosystem in general terms of a quality and organic viability superior to the original one. What I am saying is that when man interferes the consequence should be an improvement on the original position.

The United Nations makes this clear in its definition of conservation as "the rational use of the earth's resources to achieve the highest quality of living for mankind". This definition would be equally valid for economic development.

Man's failure to comprehend and apply this fundamental principle as an indispensable condition of development has wrought incalculable waste in the natural wealth of the earth and perpetrated a devastation upon life and nature, itself, which defeats the imagination. The gravest and most compelling aspect of this fact is that carnage has grown in scale with the growth of the human ingenuity which has obviously outpaced human wisdom and sensibility. As a species we behave sometimes like a little boy with a gun.

The view that I hold of the ecological movement is that wisdom will march with the planning and development.

I would now like to give some examples of the global manifestations of this failure and its consequences.

It is well known that crops of the grass family especially the sugar cane, maintain and enhance the humus content of soil and thus sustain the organic elements such as earthworms and micro-organisms in the soil. They have this effect because a large proportion of the foliage remains on the soil and is decomposed into humus. On steep lands they have the added advantage of binding the soil with fine fibrous roots and thus prevent sheet and gully erosion by rain. In this case the environment is enhanced. Where ginger is grown on slopes it tends to have the opposite effect since ginger is a tuber and it is reaped by the removal of the roots, thus loosening surface soil and making it liable not merely to be eroded by floodwater, but by "flowing" downhill by itself in dry weather.

In a rather different sphere, the diminution of the whale population diminishes the life inhabitants of regions far distant from the natural habitat of that great creature. Today it is conceivable that the salmon might disappear because of the degradation of its spawning grounds or the pollution of the currents and streams by which it reaches them. Trauma is known to have been inflicted on species of terrestrial, as well as, marine life through pollution of the sea. In this age of super-tankers, how much shipping can the marine life of a given sea tolerate?

Oil tankers appear to be a particular menace to Marine Ecology, a menace which incorporates a direct and deadly impact as well upon the terrestrial and, indeed, the entire human ecology. No audience in Britain needs to be reminded of the tragic drama of the Torrey Canyon. Two things must be pointed out. The first is that havoc is wrought on the same scale when a tanker is wrecked in distant waters - and they <u>are</u> wrecked in distant waters. The other is that in normal maritime transport operations there is constant bleeding of oil from tankers into the seas, the scale of which has been numerically estimated at 1.3 million metric tons per year.

According to the literature on the subject, the Baltic is dead - the Mediterranean is dying; not only are the higher species of marine life such as fish being directly poisoned, but grave attrition is being continually inflicted on plankton which is a crucial link in not merely the marine but in the entire world food chain. Bleeding could, I understand, be stopped or at least considerably minimized simply by imposing stricter standards of tanker construction, seamanship and of marine transport management as a whole.

I know that the attention of the International Maritime Consultative Organisation has been drawn to this grave matter and I would like to believe that UNEP and CHEC as well as national governments themselves are pressing for international action to avert this clear, present and dire threat to all life on this planet. The current Conference on the Law of the Sea would be worthwhile if only it produced and secured international acceptance of salutary and enforceable legislation providing against degradation of Marine Ecology anywhere in the world, the decimation of threatened marine species and the contamination of coastal areas.

Turning to our more local experiences, one of the baleful forms of human interference with the Jamaican ecology is urbanisation, which, of course, includes suburbanisation. The ecological penalty of this form of development in Jamaica is that it is concentrated predominantly in one area - Greater Kingston. The population of Kingston rose from approximately 250,000 in 1943 to something like 750,000 in 1973. The sum of the populations of all other urban centres in Jamaica remain less than 250,000 today. Kingston is virtually the sole repository of all the essential requirements of civilised life such as, public administration, finance, trade, industry, medical and other social services, and inevitably, of opportunity for employment and personal growth and advancement. As the broad demographic statistics demonstrate, this growth centre incorporates the inherent power of magnifying itself. This centre grows because all the services and all the amenities and all the opportunities are in Kingston - everybody comes to Kingston. And so there are more opportunities and more amenities in Kingston so more people come to Kingston and Kingston grows larger. And because it grows larger more people come and it grows larger still! And with each scale, with each measure of growth, the establishment of the amenities of civilized life becomes less likely, anywhere else in Jamaica. And that is a very, very bad thing indeed. I say this as a man and as a Jamaican. I am glad to say that we have become aware of this in Jamaica. And my Ministry and other Ministries within the government are thinking of means by which we can effect a more practical and humane distribution of urban facilities in the island so that other people can grow up and achieve the fulfilment of their natural endowments without having to come to Kingston or having to come to London.

The Kingston Metropolis is itself sited on splendid agricultural land in a country in which the ratio of arable land to population is low, where population growth is high and where there is already an intolerable deficit in food and other agricultural commodities. Kingston is now in actual competition with the adjacent agricultural areas for water. This might be just as well since, if present trends are tolerated, the city will swallow these lands themselves eventually. Kingston is now consuming the best part of three rivers while overdrawing its own ground water resources. It has virtually reduced its harbour, one of the finest and most beautiful in the world to a polluted pond by making it a catchment for waste and is now beginning to poison its own subterranean aquifers in the same way.

It is now stripping the hills, in whose lap it rests, of vegetative cover, so impairing their beauty, and also destroying in the process considerable additional ecological assets such as bird life and underground water resources.

The Kingston model is being duplicated, fortunately on only a comparatively minor scale, in other places in which urbanisation and human settlement are taking place. The evil in the Jamaican model of urban development resides not merely in the spoilage and degradation of its own immediate habitat and the impoverishment of the entire human ecology of the country at large, but that it exacts an economic and social cost beyond what a country such as this can ever conceivably be able to meet.

One of the basic features of this system is the desertion of the central city for the freshly created suburb. I will term it "the sprawl system". Urban or rather surburban sprawl not only consumes many times the acreage required by high density development thus inflating the price of land by creating scarcity, but increases or rather multiplies the cost of such infrastructural amenities as roads, water, sewage, electricity in terms of land, material, establishment and maintenance. But the penalty goes much further. It includes massive increments in the volume and expense of travel and transport, and lethal side effects such as air, water and soil pollution, traffic density, energy costs and so forth.

This system of urban development must clearly be arrested, modified and reversed. A rational distribution of urbanisation and indeed, human settlement, will have to be evolved as a deliberate function of national policy. This process must have a frame of reference including such factors as proximity to natural economic resources, social and ecological cost, the general impact on a small island such as Jamaica.

Turning to problems of marine ecosystems, Kingston Harbour is a semi-enclosed lagoon with a low rate of water exchange with the open ocean. Many rivers, streams and gullies flow into it which carry nutrient materials from the land. The catchment area of the harbour includes much productive agricultural land on which fertilizers and pesticides are used and some of these ultimately find their way into the harbour. There are large numbers of industrial plants located on the shoreline of the harbour. Secondary treated industrial effluents from at least fifteen plants enter the harbour. This means that there are large quantities of nutrients discharged. Moreover, untreated industrial effluents from many small plants also enter the harbour through gullies and small streams. The two main municipal sewage plants discharge only primary treated sewage which means that the solids have been removed but large quantities of nutrients and dissolved organic matter are released into the harbour. Garbage and miscellaneous waste products are also dumped along the shoreline and this also contributes to the pollution of the harbour. Recently, with increasing numbers of ships visiting Kingston Harbour, oil pollution in the form of contaminated bilge and ballast water has become noticeable.

Because of this, serious problems have developed in the harbour. The nutrient level has reached to about 60-70 times as high as in the open ocean and this has led to a very high level of primary productivity. This has led to the persistent occurrence of 'Red-Tides' throughout the harbour. Red-Tides are due to a sudden increase in the population of single-celled microscopic plant organisms. The frequent occurrences of Red-Tide is usually due to imbalances in the supply of chemical nutrients and is a clear sign of instability of the biological processes in the harbour. These phenomena have caused oxygen depletion in the harbour water with consequent repeated fish kills.

In the bottom of the harbour remarkable changes have occurred in the past 7 years due to a rapid decline in the amount of oxygen available. Changes have been noted in the marine animals which inhabit the bottom muds. There has been a rapid decline in the variety and in the absolute number of animals. These changes are due to the decline in the amount of oxygen available in the deeper water, the oxygen being used up mainly in the biological decomposition of organic wastes accumulating in the water and in the bottom of the harbour.

The harbour is a complex natural system with a very delicate ecological balance. It has been further disturbed by the building of the causeway and major land reclamation activities like Newport West which have probably significantly affected the equilibrium of the ecosystem of the harbour by altering the water circulation and mixing pattern. These types of physical alterations of the harbour may have accelerated silting of the harbour by altering natural current patterns.

The situation has also been aggravated by the destruction of the mangrove swamps along the shoreline of the harbour for the purpose of reclaiming large areas of land for real estate development. These swamps used to serve as highly productive areas which generated vast quantities of organic materials and thus contributed to the food chain in the harbour. The swamps used to protect the shoreline by establishing a barrier between the shoreline and the harbour. They also served as sheltered areas for marine organisms to breed and for the larvae and juveniles to feed on high quality organic nutrients.

Oxygen deficiency in the harbour water caused by organic pollution and high primary productivity, together with man-made physical alterations such as the building of the causeway and the reclamation of the mangrove swamps have altered the overall ecology of the harbour to such an extent that where in 1967 there were commercial quantities of fish and shrimps in the harbour, today there are almost no fish or shrimps. One of the largest natural harbours in the world, which once served as a commercial fishing ground and as a superb recreation ground, has been reduced to a large body of polluted water because of indiscriminate human interference.

At Negril, the obvious showpiece is Long Beach, seven miles of white coral-sand beach. Its attractiveness to tourists has made it a prime target for recreational development. But such an unconsolidated deposit is physically and ecologically fragile. The sandy beach is being continually shaped by the sea. Wave action brings new material onto the shore, along which currents carry it in the direction of the prevailing wind. Interference with this longshore drifting by human structures disrupts the continuity of the process, leading to leeward erosion and the eventual undermining of the structures themselves. If, as often happens, sand is removed from the beaches for the purpose of construction elsewhere, beach stability is reduced and the threat to human coastal structures increased in times of storm and high seas. Similarly, disturbances of stabilising dune vegetation makes the strand less resistant to changes wrought by wind and wave. In the lee of the dunes, another problem follows overdevelopment - extraction of too much fresh water from the underlying sands invites saltwater intrusion and possibly the debilitation or death of the native organic community.

But it is not enough to consider just Long Beach. The coral strand cannot be treated in isolation from the whole coastal ecosystem of which it forms a part. Behind the beach lies the Great Morass, some 6,000 acres of seemingly useless marsh, swamp and streams. Offshore lies a reef of living coral. Removal of these components would quickly demolish the Beach.

Of most direct importance is the fringing reef, for upon this, Long Beach depends for continual supply of finely granulated reef debris formed by wave action from the calcium carbonate remains of coral and algae. A healthy reef is one of clear water and strong surf. A reef with diminished growth rates is one with diminished powers of recuperation and growth and capacity to contribute to the structure of the Beach. The latter condition normally occurs offshore from the mouths of sediment-laden rivers, where fine silt carried in suspension makes the water too turbid for optimal reef growth, and dissolved oxygen and salinity may fluctuate widely with changes in the amount of runoff. Reef survival often depends on reciprocal action between marine and terrestrial factors. At Negril the reef survives because of the morass, which detains and filters floodwater before releasing it gently to the sea. Drainage or filling of the morass for agriculture or settlement would destroy the system's buffering capacity, leading to impairment or destruction of the fringing reef.

So far, the Negril coastal ecosystem has been discussed primarily from a physical and structural point of view. Of equal significance, however, is the ecosystem's biological function. Coastal wetlands like the Great Morass are among the world's most productive areas. Nutrients washed downstream are held and stirred by the tide near the river mouth. These contribute to the vigorous growth of mangroves, grasses and other plants, which in decay become detritus fodder for myriad micro-organisms. The rate of production at this, the lowest level of the food chain is so rapid as to support a large biomass of dependent organisms, from invertebrates like shrimps and shellfish to the majority of native game and food fishes at still higher levels of consumption. Many fish species begin life in estuarine conditions to spend adulthood elsewhere in the reef or the open sea; some species visit only to feed; others again live most of their life cycle in or close to the wetland. It is clear that human interference with the Great Morass can reduce its value not only to commercial and subsistence fishermen but also to the nature enthusiast on the reef. Needless to say, land use practices involving local concentrations of herbicide, pesticide or other toxins in the runoff have, in the long or short run, devastating effect on marine productivity.

The seasonal nature of the coastal ecosystem must not be overlooked. This is especially noticeable in the case of bird life. Some species make an appearance only for a month or so to breed and then depart. Others spend the winter, or parts of the year when passing through to points farther north or south. Like the coastal habitats themselves, many of these species are becoming scarce. The coastal ecosystem is thus being diminished and degraded.

As has been remarked elsewhere in this Paper, the massive intervention of man in an ecosystem may be justified where it can be proven that the system's natural productivity is either capable of significant enhancement or is insignificant to begin with. Where productivity can be increased, some form of agricultural enterprise may begin; where productivity hardly counts, human settlement may take place. Neither kind of development is justified at Negril. Clearly, all plans for the further development of this fragile, dynamic and productive ecosystem must be subject to the most stringent regulations.

Agriculture can reasonably be regarded as the original

and major spoiler of the Jamaican ecology. It can be represented as having begun on the one hand with the establishment of the Sugar Industry on the alluvial plains which run between the hills and the sea on the south of the island and in much smaller parcels on the north also marching with the sea. It began more or less concurrently in the hills with the establishment of the Coffee Industry, where the soil in terms of physical and organic composition on the one hand and climate, including altitude, temperature, humidity and light on the other constituted quite literally the ideal habitat for that crop. Unfortunately, in those days the highlands were the only hospitable residential location for European settlers and even the senior employees on the sugar estates used to live in the hills. The plains were quite literally pestilential for Europeans in those days. These highlands were quite predictably less fertile generally than the plains, and before and after the Emancipation slaves were allotted lands there and cultivated them in food crops for their own subsistence.

In lowlands and highlands alike the ecology was subjected to savage and thoughtless abuse. In both cases vegetative cover was stripped by continuous clear-cutting on a vast scale. This process destroyed the shelter and the food that sustained birds and other wild life, and of course, the species varied with elevation and climatic factors. The decimation of flora, including rare timber as well as delicate and beautiful exotic plants, occurred on the same scale. Sheltered horticultural stands were explicity if thoughtlessly ravaged for convenient utilitarian purposes. Coasts and river banks alike have since continued to be stripped of protective tree growth to the detriment of land, water, flora and fauna alike. It would require deep and extensive research to identify all the affected species, terrestrial and marine, not to mention soil and water themselves. I, myself, am sustained by the knowledge that these losses need not be final and irrevocable, in their totality at any rate - and by the belief that an informed and sensitive community can in time substantially restore these losses and, perhaps, by judicious transplantation of species restore the natural ecology of Jamaica into a comparable balance and a comparable balanced opulence.

Jamaica is an island. It is a small island. It is already an overloaded boat. We have disregarded that, my own generation and our predecessors. We think that we can emulate larger countries, that we have the same sort of cities as are established in the United States where the population is nineteen to the square mile. In Jamaica you cannot travel in a straight line for more than twenty-two miles before reaching the sea. Obviously, we have to act more thriftily in our use of land than say Britain, which is, incidentally, regarded nowadays as a small island. We cannot allow human settlement alone to gulp large portions of the available land. Or we could reach a point ultimately at which we can accommodate nothing but human settlement. It could happen.

And so we have to feed into economic and physical planning, and into ecological husbandry, considerations of the space at our disposal and the ecological characteristics of that space. Because it is small, because it is an island, the ecology is fragile. In considering the location of a significant development complex involving the use of coastal areas, we must realise that coastal strips could devastate invaluable resources of marine as well as terrestrial life. People who plan and develop should know of the existence of the sort of creatures which are invisible but which are there and may be vital to us. It is something of consequence to a small country if it interferes in such a way as to destroy a mile or two of the breeding ground of a fish species, as has been done. An accident resulting in pollution, causing the overflow of poisonous matter into the sea can deprive Jamaica of millions of fish, a problem that might take years to overcome.

We therefore have to devise a process of development and a set of criteria for development which have full regard for the ecological constraints under which we operate. There is nothing essentially in conflict between good ecology and good development. They are the same. And they should be planned for at the same time. And our own idea in Jamaica is to make ecological husbandry and development planning into one single integral process.

In Jamaica we are awakening to the challenge of the essentially murderous and suicidal character of our attitude to our own natural heritage. Our grasp of the fact that the crisis is total and can only be solved integrally is quickening.

Successive governments had been persuaded to accord perfunctory recognition to the fact that soil erosion should be arrested; that birds should be protected; that beaches should be brought into the national custody and control for the public good. Organisations were set up in each case and in each case maintained at the level of sheer survival in proof of good faith.

At the birth of the present government, however, two significant events occurred. The responsibility for the public health aspects of environmental control was vested by extension of the title of Ministry of Health; and the Ministry of Mining and Natural Resources was created with the broad function of resource management in terms of surveys, investigation and research, planning, development and conservation, which latter includes restoration and rehabilitation.

All bodies previously concerned with the functions of resource conservation have now been merged into a new agency called the Natural Resource Conservation Authority which is now in its formative stage. It administers existing conservation legislation which has proven to be adequate for initial requirements; but the incipient Authority is now engaged in developing an ample functional establishment as well as the necessary legislation which, together with the disciplinary resources and organisational and other apparatus at the Agency's disposal, will enable it to go meaningfully into the resuscitation and enhancement of the Jamaican ecology and contribute meaningfully through international fellowship and participation in the pursuit of this vital and noble task in the world at large.

Returning, now, to this more global outlook, I would call the ecological movement the revolution of the 20th Century. Many people look to independence movements, to the advent of atomic science and of electronics and the shrinking of the world by technological improvements of one kind or another. But unless these improvements, unless man's management, man's husbandry of this earth and his use of science and technology, are carried out with due regard to their effect on the society of this earth then it is inevitable that our chances of survival will eventually be that much slimmer. One of the things I have observed for many years is the distance by which mere intellect, mere knowledge, the power to think, the power to understand and comprehend intricate systems have outpaced what I call wisdom, which is that understanding by which man's actions are tempered by the considerations of all the factors and of the consequences of these actions. It is that wisdom which eventually, I hope, will enable us to see that we are all on the same planet, whether we are third-world countries or first-world countries - whether we belong to the northern hemisphere or the southern hemisphere, whether we are capitalists or socialists, we are all on the same planet. We are a part of one another and this earth is our estate and we must husband it. We must manage it with due consideration to the welfare of that which we manage and I would wrap it up in essence to say our respect for life and our responsibility for life. If we can't inform this generation and succeeding generations, if we cannot inculcate this principle then we will be that much the worse off and the outlook will be bleak indeed.

NATURAL RESOURCES CONSERVATION AUTHORITY AND DEPARTMENT

1 The government has decided to create a Natural Resources Conservation Authority which will not only absorb the activities of certain existing organizations and statutory bodies under the Ministry of Mining and Natural Resources but will also develop the capabilities required to deal, in the short as well as the long term, with ecological problems affecting the welfare of the country.

2 The Island is at present in the grips of a prolonged drought and in the Corporate Area it has become necessary to institute stringent measures to stretch dwindling supplies of water. The problem is not only one of harnessing water and delivering to consumers but also of conserving present resources and restoring them by sound resource management practice to their former abundance and natural quality.

3 Very recently, public attention has been focussed on the problem of ecological damage in the bauxite areas arising from:

- a) the intrusion of caustic soda into the Magotty River and the consequent damage to fish and other aquatic life;
- b) seepage from red mud lakes into underground aquifers and,
- c) the release of harmful dusts from mining, manufacturing and processing activities into the atmosphere.

4 Sewage and industrial wastes entering our streams, rivers and harbours constitute a major problem - levels of tolerance can be determined but can only be enforced through constant and effective monitoring. This is impossible at the present time because of the total absence of the necessary apparatus. The condition of the Kingston Harbour continues to degenerate because of the influx of poisonous and corrosive waste. There is general agreement that urgent action is required to minimise and reverse this deadly process. The Harbour itself is a complex natural system which

operates by way of a very delicate ecological balance between many organic factors. Major land reclamation and connected activities are bound to exert harmful effect on the natural equilibrium of this system. The productivity of the harbour has already been reduced by the destruction of mangrove areas and the alteration of the siltation pattern by the disruption of natural current patterns.

5 The erosion of our beaches and foreshore is generally recognised as a growing problem. In recent months there has been severe beach erosion in the areas of Doctors Cave and Heritage Beach in Montego Bay. These are due very largely to human intervention in the interaction of complex natural phenomena which control the rate and pattern of sand transportation in this section of the coast. Comprehensive oceanographic studies of this and other areas are required in order to identify causes and design remedial measures.

6 Two oil spills which occurred last year - one in the Portland Ridge area, Clarendon, off Portland Cay, and the other in the Salt River Clarendon area, have highlighted the need for closer scrutiny and regulation of fuel oil handling and distribution in Jamaica. It is clear that both were done deliberately - one by a departing oil tanker and the other by a land-based carrier. The consequential destruction of fish and other wild life as a result of both was enormous.

7 Changes in the natural environment pose a threat to the continued existence of many species of wildlife. Many have already been exterminated. The Manatee, the crocodile, exotic birds and fish fall prey to current malpractice of over-hunting and wanton destruction of breeding grounds and feeding areas. Illegal hunting of crocodile, dynamiting of fish, filling swamps, clearing forested areas, the destruction of ferns and orchids as well as the casual disposal of sewage and trade effluents are but a few examples of the indiscriminate treatment of the Island's natural environment.

8 The projected Luana Industrial Complex and aluminium smelter have the potential to provide massive material rewards for the country. However, such activities as the construction and operation of industrial facilities sometimes result in losses of natural assets which in the long run generate a net liability to the country. Such losses,

which vary widely in magnitude may result from the failure to adequately consider environmental consequences during project planning and design or from lack of knowledge and information necessary to determine the eventual impact.

9 Industrial development inevitably produces large quantities of liquids, solids and gaseous wastes. The intrusion of these wastes alter the environment unless their effective control is planned into the location and design of the relevant operations.

10 The increasing impact of man's developmental activities on the natural environment demands immediate action to minimize the negative aspects by incorporating resource management considerations in the planning and control of development. Hillsides need not be ravished by poorly planned housing developments, the indiscriminate clearing of trees or wanton agricultural practices. The ecology of the Island must be protected from man's increased ability to alter the face of the land.

11 It is necessary therefore to institute a Development Control Procedure by which the normal functions of the Town and Country Planning authorities will be strengthened by the addition of ecological skills to both the planning and control of the nation's physical development. This process will require the establishment of a data bank to store and distribute data on the natural resources of the Island. An information exchange facility to provide up to date knowledge of world-wide developments in techniques for the conservation and development of natural resources is also required.

12 The foregoing is an attempt to draw attention to certain ecological problems facing the country. What is required now is to consider the institutional and technical capabilities which the government must establish in order to deal with these problems. The existing agencies at the disposal of the Ministry of Mining and Natural Resources are:-

the Beach Control Authority
the Watersheds Protection Commission
the Wildlife Protection Committee
the Natural Resource Planning Unit
the Marine Advisory Committee
the Kingston Harbour Quality Monitoring Committee

13 It is proposed to amalgamate all these agencies into a strengthened and expanded institution to be known as the Natural Resources Conservation Authority and pending the introduction of legislation the "Authority" will operate by means of a common membership on all existing bodies and organizations.

14 The enabling legislation will among other things, create and constitute the Natural Resources Conservation Authority and charge it, within the framework of the proposed Act, to advise and help the Minister:-

- to increase public understanding of the Island's ecological systems and promote methods for the conservation and development of its natural resources;
- to raise the quality of life by increasing the public awareness of the natural beauty of the Island and widening the availability and accessibility of outdoor recreational facilities;
- to determine policy to be followed and standards to be maintained in the management of the Island's resources of land, water, air, flora and fauna in the interest of the present and future generations of Jamaica;
- to promote and ensure the wise use of the nation's natural resources by the establishment of an ecological review procedure for all relevant development proposals;
- to implement programmes for the conservation and development of natural resources;
- to collect, store and distribute, data and information on the development and conservation of the Island's natural resources.

The Authority will be serviced and assisted by a Civil Service staff to be organized under a new Department known as the Natural Resources Conservation Department comprising five Divisions:-

Recreation and Conservation
Watersheds Engineering
Aquatic Resources
Resource Management; and
Administration

Divisional Directors will be responsible to a Principal Director.

16 <u>Recreation and Conservation Division</u> The purpose of this Division is to plan, develop and manage where necessary, National Parks, Beaches and unique terrestrial and marine Beauty Spots as well as Wild Life and Ecological Preserve

Areas for the preservation of unique species. It will be generally responsible for the conservation of land, species of animal and plant life as well as for the provision of recreational facilities.

17 Watersheds Engineering Division This Division is intended to determine the need for, and to undertake, water conservation through water management projects. These projects include Upland Watershed Management, River Control, Flood Control and Sea Control. This engineering-oriented Division will plan, design and maintain all such projects on an island-wide basis. The major projects will continue to be undertaken on an agency basis by such agencies as the Ministry of Works and Local Government Bodies which presently have the capability to undertake such projects.

18 Aquatic Resources Division The Aquatic Resources Division will be responsible for surveys, investigation and monitoring of water quality, beaches, seabeds and wetlands. This Division will utilize skills in chemistry, aquatic biology and physical oceanography to study the physical and biological stability of our coastal resources as well as to monitor the quality of our inland and coastal waters.

19 Resource Management Division This Division will be organized in two branches, with functions allocated as follows:-

Ecology Branch This Branch will contain specialist skills in ecology and will work closely with the Physical Planning Division and the Local Authorities in the application of an Ecological Review Procedure for providing the necessary ecological inputs to the planning and development control processes.

Data Branch This Branch will be responsible for the storage, processing and distribution of natural resources data collected by the Authority. It will also maintain an updated reference to all relevant data collected by other agencies of government and international organizations.

The Data Branch will also be responsible for the Public Education Programme and Public Involvement Procedures of the Authority.

20 Administration Division The Administration Division will be concerned with administrative matters such as finance, personnel, registry, etc., which are crucial to

the efficient performance of the Department. It will also be responsible for training which will be especially vital in the early stages of the development of the institution.

21 A total provision of $600,334 has been included in the 1975/76 Budget now before the House, to finance the operation of the organisations mentioned at paragraph 12 and which are to be eventually absorbed by the Natural Resources Conservation Authority and Department.

22 A functional profile of the proposed Natural Resources Conservation Authority and Department is attached together with detailed estimates of staff and other requirements for the new Natural Resources Conservation Department during the financial year 1975/76 amounting to $600,334.00.

23 I will move a resolution in due course requesting the House:-

a) to approve the establishment of the new Organization and staff as listed;
b) to agree that the posts may be filled pending provision of funds for the new Organization - the Natural Resources Conservation Department in the First Supplementary Estimates; and,
c) to note that pending approval of the First Supplementary Estimates the new Organization would be financed by means of advances from the provision of $600,334 mentioned above.

Editorial note: The profile and further information on the Natural Resources Conservation Authority and Department is available from the Ministry of Mining and Natural Resources, P O Box 223, Kingston, Jamaica.

HUMAN ECOLOGY IN THE ASIAN ENVIRONMENT

F Z KUTENA

I shall be speaking on a subject which was suggested to me by the Secretary-General of the Commonwealth Human Ecology Council. It is called: "Looking into the Asian Scene in a Commonwealth and International Setting". When one of the members of CHEC saw the cable confirming my willingness to accept the challenge, he remarked: "His subject matter covers all things to all men especially all CHEC men". He was right.

Since 1960 I have served in eleven countries and since my graduation in engineering in 1939/40 I have lived in seventeen countries in Europe, Australasia, Asia, Africa and Latin America. Perhaps this experience and exposure to so many cultural, social and economic systems, ranging from Nazism, Fascism, Communism, Colonialism, Democracy and all the different -isms of developing countries, prompted the Secretary-General to consider me suitable for this task.

The subject matter covers everything and becomes unmanageable if we attempt to carry out detailed research on Asia and synthesize the findings. I did it and I found the text dry and uninteresting.

It is not the purpose of this lecture to go into all historical details underlying the present pattern of Asia. I shall touch on various countries, with more emphasis on Bangladesh, because it represents all the problems Asian countries are facing in various shades and combinations.

Asia, which covers one-third of the land surface of the earth has been influenced by three major civilizations: the Chinese, Hindu and the Islamic. According to Steedman in "The Myth of Asia", 'Asia as a social unit has no existence at all'. It has however produced complex and rich civilizations. It is marked by contrasts and diversity, in language, religion and culture and in the organisations and social patterns of its societies. It calls to mind the magnificence of a temple and the squalor of airless mud houses, the age old sampans of Hong Kong and the modern apartments of Tokyo, the seasonal floods of the Brahmaputra delta or the sun-scorched desert of Sind, the jargas of the Pathans or the Kingdom of Siam.

The great thrust outwards by Western powers in the 19th century brought new technology, alien governments and economic interests which transformed certain parts of Asia. The end of World War II brought with it the dismemberment of the Colonial framework. Asia was moving. It was alive with ideas and aspirations. In the short period of 30 years it has moulded governments ranging from an ideologically severe form of Communism to Western-style Democracy. It has had to grapple with crushing economic problems, with great poverty, shortages of trained personnel, few facilities and staggering population increases. It has looked towards the industrial nations for assistance being torn between the need to finance and implement development and the suspicion that such dependence could result in economic manipulation. Parts of it are still at war, reflecting the conflict between two Western-born political philosophies.

While the Commonwealth countries in Asia have adopted democratic constitutions based on the centuries-old experience of Britain and guaranteeing basic rights to their citizens, some have offered less freedom and China has chosen Communism.

One sometimes concludes that a parliamentary system newly introduced into a society with a high percentage of illiterate people can lead to misuse of the freedom and even perpetuate and aggravate already great social differences amongst the people. The excesses of a purely profit-motivated western-style free-enterprise can produce an imbalance.

The Chinese economy on the other hand does consider profit and growth but as a consequence rather than the main objective. They claim that economic expansion will automatically follow once the society has improved its standard of living and its forces have reached a better balance. H V Henle, a Regional Information Adviser of FAO, in his recent Report on China's Agriculture has this to say: "The Chinese think that their experiment may produce a living and affirmative answer to Engels' old question of whether forms of optimum social organisation can be devised that are particularly conducive to accelerated economic growth and social well-being. Having absorbed Marx's teaching that being determines consciousness (or that the economic substructure determines the ideological superstructure), they hope to reverse the causal connection. By judiciously improving the minds of the people and the structure of society in a politically predetermined pattern they expect to better the underlying mechanisms of economic activity, with a tremendous release of latent, materially productive force as an inevitable result".

The Chinese therefore reject Western economic thinking. They claim to prove that economic growth and the creation of wealth comes from unselfish concern for the common weal and from the democratic structure of society. ('Democracy' as China understands it.)

This can be contrasted with Japan, where the welfare and education of its workers is frequently handled by the same giant corporations whose creative industrial and business ability has enabled them to surpass most of our Western industrial nations and made Japan a great trading nation.

While the Japanese are much concerned about fuel and the markets which could be adversely affected by mismanagement of the Middle East situation, the Taiwanese seem to have reached a stable economy and have the management capabilities to cope with eventual fluctuations of markets and related domestic problems.

Similarly the Republic of Singapore has made tremendous progress in all areas of economy and environmental improvement. Singapore as a member of the Commonwealth competes for banking and commerce with Hong Kong whose flexibility of adjustment to suit changing trends in the trade patterns is commendable. It seems that the economic boom for both Singapore and Hong Kong is ahead. They may become the banking centres of Asia occupying a position similar to that of Switzerland in the West.

A typical raw materials supply country with under-developed agriculture and industry is Malaysia. It is a country rich in tin, rubber, palm oil and timber which offers possibilities for local industrial development. Its climate and potential land resources can be developed to meet domestic food and fibre requirements. A CHEC 'Case Study' could perhaps be of benefit to the planners there.

Going south-east we find the country of thousands of islands, the Philippines attempting to solve its economic and political problems. Australia has indicated possible assistance with the first token recruitment of Philippinos for public works as the Common Market countries in Europe employ southern Europeans.

As with Malaysia, Indonesia has been fortunate in having an abundance of natural resources, particularly oil, which may smooth out any deficits in food and basic necessities if such a situation occurs. In the long projection great investment opportunities exist in all sectors here for Singapore's banking houses.

South of Indonesia the Australians are reforming their economy to suit their heightened role of raw material and food suppliers to the rest of the world and to Japan in particular. In human resources and cultural development we in Australia should not go 'all out' to become miners and pasturalists although in a material sense it might be economically justified. I feel we should adopt policies supporting the development of a healthy multi-sectorial society for which process we need industrial development and research as a means of achieving balanced growth for the people and the quality of their environment. Imagine for a moment what could happen if we neglect this and channel our human resources into mining and the auxiliary services. In 50 years time should the mines phase out we would be left with a society ill equipped to compete in a highly technical world. At that time Asia will be technically, culturally and scientifically at its peak. We might even see a reversal of the present situation with Asia offering training and development assistance. Australia should therefore re-evaluate its long term developmental policy, spend money on education and research and support the industrial development of small and medium sized industries.

Referring again to South-East Asia, one can recognise signs of political maturity in the leaders on both sides of the equation - Asia and the World. The Commonwealth must continue to promote a climate of co-operation, tolerance and an understanding of mutual interdependence through free exchange of knowledge and assistance in ecological research. Top on the list of research projects are Case Studies orientated towards creation of new employment opportunities for the rapidly rising numbers of unemployed or under-employed people.

Improvements can be made by changing some of the existing business patterns. Consider for example the License Merchant in a developing country. Traditionally he belongs to the local elite, is well-off and commands respect and influence. His function before independence was to export raw materials and to import goods for domestic consumption. The government regulated this process by issuing appropriate licenses. His interest today is to continue this service and he is reluctant to switch from import to local manufacturing because of political instability and labour problems, ineffective law and order situations, inefficiency in the supporting services and raw material supply systems.

CHEC could perhaps investigate some of these manufacturing fields where domestic raw materials are available and where the wealthy countries could offer assistance in building such industries on a 'turn-key' basis. This would have the added benefit of conserving necessary foreign currency reserves.

Much remains to be done on the world food front. The World Food Conference in Rome and the Population Conference in Bucharest highlighted the tasks facing the international community. It calls for a unified action programme to eradicate the evils of ignorance, famine and the like. This can be done only by combined efforts at an international level and cannot be tackled in isolation. A food production increase can be achieved if more encouragement is given to the farmer. He must be paid more per unit of his produce. This is the best motivation in Asia and I believe anywhere in the world. It will have the effect of bringing marginal land into production as farmers make additional efforts to increase their yields. In many countries it is an economic possibility if less 'middlemen' and more efficiency can be brought into the chain between producer and consumer.

Progress is slow if we think of the life of a villager and its quality improvements. Once all marginal land has been put under cultivation by traditional methods further production increases require technical know-how and additional energy inputs in varying forms - fertilizers, fuel and engines. This in turn needs initial capital which the Asian farmer usually does not have and the whole system needs efficient supporting sub-systems in the operation, maintenance and replacement of mechanical devices, storing and distribution, marketing and processing. If any one of these inputs is missing or incorrectly administered the system fails to meet the expected results, often failing completely as is the case with a breakdown of the water supply system for irrigation. The resulting crop failure is a critical thing for the farmer.

I have selected this problem for closer examination to see whether there are natural, developed, sub-systems at village level which could be utilized, with a little effort, for operation, maintenance and replacement of mechanical innovations in the villages. There are two possible approaches to this problem. One is a government service system right down to the farm level or, alternatively, a private enterprise system. In Bangladesh the former system, originally introduced, is now being re-assessed with the objective of replacing it with existing village 'Blacksmiths' workshops upgraded to a simple mechanical repairs level. This will operate on a trial basis in three districts.

As a background to this problem I should mention that in order to reach self-sufficiency in food grain by 1978, the Agriculture Development Corporation of the Ministry of Agriculture is installing 45,000 lowlift pumps, 19,000 deep tubewells and 15,000 shallow tubewells - in total some

80,000 diesel engines and 80,000 pumps. If we add to it mechanical devices already in the environment we can easily double the figure. These machines must be properly maintained and repaired, often at short notice as a timely water supply is important for irrigation.

A close examination of the village social system offered a solution - the local blacksmith or Kamar. He is both blacksmith and toolmaker, producing and repairing farm tools and other implements for the village community. There are some 60 thousand to 120 thousand Kamar/Mechanics in Bangladesh. They came into being through a process of natural selection within the village eco-system when the needs for their services developed. The demand built up slowly over centuries from very simple technology to the manufacture of knives, jute cutters, hoes, spades, axes requiring more specialised knowledge in steel forming, hardening, welding and so forth. His present workshop consists of a fire place with bellows and a set of normal tools. It is usually about 10 x 10 feet, made of grass mats and bamboo with a mud floor. His work is carried out in a sitting position on the floor. The Kamar becomes a tradesman by his own choice. As a farmer's son at the age of 10 or so he joins the local Kamar as his assistant. When he feels he has learnt all the skills of the trade he starts his own workshop. If he has attended 6 classes at the local school he can go to town and try to join one of the trade schools and advance his knowledge. This does not happen very often.

In relatively recent times mechanical innovations have started to move into the village eco-system. Bicycles, motorbikes, outboard motors, diesel pumps and cultivators all created a demand for repairs. The local Kamar began to develop new mechanical skills. When the demand was large a mechanic from an urban area might join him as a partner or move into the area and open his own workshop. Usually lack of credit, to some extent a lack of specialised knowledge and absence of public knowledge about projected demand areas offering new opportunities are the main reasons why the natural adjustment is delayed.

In order to get some <u>dynamics</u> and efficiency into a nation-wide mechanical repair workshop system four conditions must be fulfilled:

1 Private enterprise
2 Fair profit
3 Possibility of growth
4 Political stability, personal security, law and order.

An analysis of the Kamar's skills and their products conducted at the village level in their natural environment indicated their potential for performing minor repairs and periodic servicing of diesel engines and pumps subject to short term training and provision of the necessary tools. As a next step the Kamars' capabilities were tested on a 3YWA Ruston Diesel engine. An evaluation showed that an average Kamar, after receiving 15 theoretical and 30 practical training hours, is capable of diagnosing and solving 45% of all causes of engine failure listed in the Ruston manual. This plan includes a careful selection of potential Kamars and a bonus system designed to keep the irrigation pumps in running condition, fixed prices for standard repairs and replacements, credit in the form of a prefab workshop and tools and on-the-job training by a mobile training workshop.

Looking at Bangladesh with cool realism and having experienced the hunger and misery of refugee camps in post-war Europe, the favelias of Rio, the ranchos of Caracas and the sub-life of the sugarcane cutters of Pernambucco, I have found no parallel with the human suffering, hunger and total poverty of Bangladesh today. The stresses are tremendous and will increase with time. Drastic adjustments are needed in the whole society. Let us consult the data:

In synthesis, the high rate of population growth coupled with the inability of the government to activate and channel the people's energy and capabilities into productive activities is the core of the national economic problem. In terms of food production and water requirements, an increase in gross cultivable area from a present 30 million gross acres (net 22 million acres) to approximately 36 million gross acres is going to be required within the next 10 years or so. Solutions include improvement of the production efficiency of existing projects, introduction of high yielding varieties, and irrigation, flood control and improvement of the yields of aus and aman grown under rain-fed conditions.

Family planning will have a slow impact. We can pursue the strategies of labour intensive agriculture, which will distribute the surplus population more evenly over the country thus reducing population pressure in urban areas or, alternatively, absorb surplus people from the rural areas through the development of industries. The former approach seems to offer a better solution, considering the shortage of managerial abilities, entrepreneurship and skilled workers essential for industrial development and, one cannot omit, the danger of unrest when dissatisfied people are concentrated at one point. This does not mean to neglect the industrial sector but it does mean to prepare conditions which will help the nation to survive a very critical period with the least possible stresses. This policy would move the society in the general direction of labour extensive and intensive agriculture coupled with the development of local resources-based industries.

The Commonwealth Human Ecology Council in Bangladesh has started to prepare a Case Study of a typical rural environment which it is hoped will throw more light on the rural scene in Bangladesh and so help to accelerate a healthy and balanced growth within the village. It also started a self-help programme in a small village next to Dacca with the objective to study various motivation methods.

Bangladesh is an old, established settlement region, probably best defined as an upward transitional area where the predominantly rural economy is still in a state of stagnation, supplying the bulk of the migrant workers to other more dynamic regions. Its problem arises or is associated with obsolescence and with over-population relative to existing development possibilities. Present population density of rural areas has been estimated to be of the order of 3.4 persons per net cultivated acre as compared with India with only 1.8 or in other terms 0.3 acre per person in Bangladesh compared with 1.5 acres per person in the United States.

During the detailed planning phase it is essential first to make clear the process of re-ordering spacial relations which occur in growth situations. It is essential also to clarify the dynamic influence of spacial patterns on growth and ecological factors on which the existence of the people and of the national economy depends. Once this has been done an attempt must be made to influence the total growth pattern by matching specialised sub-regional advantages such as the supply of local inputs, ground or surface water, power, gas, clay, minerals, land, forests etc with the demand for specialised industrial and agricultural requirements. This process will not take place by itself and substantial changes in the socio-political and income pattern will have to be included.

Water resources development will take place within this overall planning concept, with the terminal objective, as mentioned earlier, being increased food production as a target. There is an unpredictability about the monsoon flooding on which the bulk food grain production depends. This needs to be taken into consideration in planning water resources for food self-sufficiency. In addition to the Aus and Aman crop seasons full utilization of the dry season must be made to grow Boro under irrigation. This can only be achieved by the use of both types of water resources available during the dry season (December to May) that is local water and 'imported' water coming to Bangladesh in some 42 streams across the political border from India. Consideration must also be given to specific ecological and environmental constraints in the individual localities. This means accepting now that certain water uses; domestic, livestock, recreational, fish or one can say, the life supporting water will always be better served from local water stored in depressions or underground often because of health reasons and because it carries the biotic stock. These water requirements will progressively consume most of the locally available water resources while the bulk consumer, irrigation, will have to depend more and more on 'Indian Water'. Before implementation of the development of water coming from India can be fully realized a necessary water agreement must be reached between India and Bangladesh.

As Bangladesh is to be found at the bottom of the great Ganges, Brahmaputra, Meghna River system, much of which begins in Tibet or Nepal and crosses India, sufficient flow needs to be guaranteed for Bangladesh. This would involve international agreements for the protection of water rights and it is a field in which the Commonwealth Secretariat could help those countries concerned to find suitable solutions.

The evening before I left Dacca a young Bengali woman friend of ours came to my home. I asked her what she would like me to add to my address. She sat down and wrote six impassioned pages which I have had to condense to a few paragraphs but which I feel have relevance as an expression of the attitudes of educated young Bengali women today.

She felt that one of the greatest obstacles to development in Bangladesh was the traditional role still being forced upon women. She was using 'development' in the widest ecological sense. I quote from her paper: "given this dissatisfaction for the first time in history with their traditional role as sex objects and chattels, it would indeed be an appropriate moment to introduce new thoughts without hurting the prejudice and pride of menfolk. The Commonwealth Society can help immensely by persuading the government of population-ridden countries like Bangladesh to organise a Ministry of Women's Affairs and thereby give a helping hand to already conscious women to be vocal and effective in a still hostile environment". She has made some other interesting suggestions; that slogans supporting family planning and respect for women should be printed on the ever-present ration cards, that family planning and the danger of excessive population should not only be aired on the media but should be instilled into all children from kindergarten age onwards throughout their school lives, that monogamy should be made the law and that free and legal abortion should be popularised.

In conclusion I should like to mention some thoughts I have about the Commonwealth Secretariat and ways in which it could increase its assistance and strengthen the ties between the member countries:

It could become a type of 'troubleshooter' when mediation or co-ordination is required between member countries as for instance in the case of disputes between countries sharing common streamflows such as India, Bangladesh and

Nepal, or the countries around Lake Victoria etc.

It could aim at development of a general economic policy which would function as an indicator for the Commonwealth countries in their decisions concerning development of their industries and agriculture.

It could perhaps promote and strengthen cultural activities.

Assistance could also be given to member countries to develop appropriate administrative systems and procedures and train public administrators.

It could also function as a centre where those countries with problems of a political, international or policy nature should seek advice, knowing that they can draw upon a pool of highly qualified specialists from the Secretariat or other member countries.

Internationally the Commonwealth must aim to break down the hard divisions into which the world has set and bridge the gap between the eastern and Communist blocs so that genuine co-existence is possible.

"THE ECOLOGICAL STUDY OF HUMAN SETTLEMENTS - LESSONS FROM THE HONG KONG HUMAN ECOLOGY PROGRAMME"

STEPHEN BOYDEN

I would like to begin by expressing my special pleasure at being invited to participate in this important series of lectures of the Commonwealth Human Ecology Council, because, were it not for CHEC, the main topic of my talk would not exist.

Three years and one month ago, in April 1972, a CHEC conference on Human Ecology, to which I was invited, was organised in Hong Kong. It was during the course of this meeting that the idea emerged of attempting a holistic ecological study of Hong Kong, involving the active participation of people from many different academic disciplines and including groups in Hong Kong as well as the Urban Biology Group at the Australian National University. Several speakers at the conference pointed out the special advantages that Hong Kong offered for a pilot study of this kind.

During the ensuing months the feasibility of mounting an ecological study of Hong Kong was further explored, and many discussions were held between the A.N.U. group and interested people in Hong Kong. Although many people, both in Hong Kong and in Australia, were highly sceptical about the whole idea, the firm decision was made at the end of 1972 to go ahead with the proposal and the Hong Kong Human Ecology Programme was born. Since that time we have learned some important lessons relevant to the general approach to and organisation of integrative ecological studies of human settlements, and we feel that we should communicate our experience in this regard, in the hope that it may be useful to others who may be planning similar studies on human settlements. For this reason I intend to talk mainly about the organisation of the Programme rather than about the results, which are in any case still incomplete.

First, I should point out that all of us centrally involved in the Programme are complete novices, in that none of us has previously been involved in field studies on human populations. This fact has no doubt contributed to some of the mistakes which have been made. However, in spite of its various weaknesses, we can say that at the present time, due mainly to much good fortune in the personalities who have become involved in the Programme in various ways, progress has exceeded our wildest hopes of 1972. For example, in the early stages, four people in Hong Kong had an immense and beneficial impact on the Programme. They were, Professor Ambrose King at the Chinese University of Hong Kong, Professor Frank King at the University of Hong Kong, Mr K. Topley, the Director of the Government Department of Census and Statistics, and Mr G. Barnes, the Environmental Secretary in the Hong Kong Government.

The broad aims and objectives of the Programme may be defined as follows:- (1) To improve knowledge and understanding of the causal relationships between the environment and living conditions of people in Hong Kong and their state of health and well-being. (2) To improve knowledge and understanding of Hong Kong as an urban ecosystem, in terms especially of the patterns of flow and utilisation of energy and of certain important materials. (3) To improve understanding of the relationship between the ecological characteristic of the urban ecosystem as a whole and the state of health and well-being of the population. (4) To improve knowledge and understanding of the cultural processes of adaptation to detrimental environmental influences.

One aspect of the Programme which we believe to be of great importance has been the existence of a central Integrating Group, made up of a small number of individuals whose main academic interest is in integrative scholarship. This core of integraters is concerned with the integration of relevant knowledge from different fields of specialism as it can collectively contribute to the understanding of human situations and problems. It is comprised of members of the Urban Biology Group at A.N.U., but includes some research assistants and a secretary appointed in Hong Kong.

Also involved in the Programme are a number of Specialist Groups made up of members of the staff of the two universities in Hong Kong and also, working as consultants to the Programme, members of the staff of the Commonwealth Scientific Industrial Research Organisation (CSIRO) in Australia.

The main work of the Integrating Group in the early stages was to develop a conceptual framework for the Programme and to invite participation of the various Specialist Groups. The Integrating Group has also collected a great deal of information to be used for the preparation of an integrated ecological description of Hong Kong, to be written in terms of the conceptual framework of the Programme. This information has come from a wide variety of sources, including government departments, academic literature and the various Specialist Groups. Originally it had been our intention to base the integrated description entirely on existing data acquired from these sources. However, because our conceptual approach was somewhat different from that of other investigators, it soon became apparent that some questions which seemed important to us had never been asked in earlier more specialised studies. It was therefore decided to include in the work of the Integrating Group a large "Biosocial Survey", involving an interview-questionnaire, aimed at filling some of the gaps in our knowledge relevant to interactions between environmental and life style variables and health and well-being. This "Biosocial Survey", which was carried out with the help and collaboration of the Social Research Centre, Chinese University of Hong Kong, and which involved almost 4,000 respondents has been completed and the results are now being analysed.

I cannot overemphasize the importance we attach to the role of the central group of full-time "integraters" in this kind of study. Integrative scholarship is difficult, and to be successful it requires a great deal of concentrated intellectual effort. It is much easier to launch a good specialist than a good integrater, and much easier to launch a good specialist programme than a good integrative programme. So many "interdisciplinary" or "multidisciplinary" programmes, both in education and in research, have been failures, because they have been set up and organised by committees of specialists who, after their occasional meetings, return at once to the relative security of the various disciplines to which they owe their primary allegiance.

Nevertheless, the role of the Specialist Groups in the Hong Kong study is, of course, equally essential. Some of these groups are working quite closely with the Integrating Group, and others are almost entirely independent. Each of these groups aims to produce a specialist report or monograph on an aspect of the ecology of Hong Kong. The subjects to be covered include aspects of the energy flow, air pollution and artificial heat production, nutrient flow, water resources and water pollution, mortality patterns, fertility patterns, noise levels and their effects, home medicine, and environment and child growth. It is intended that these monographs will be written in terms of the conceptual framework of the Programme as a whole. Thus, the participating Specialist Groups, which are interacting with one another and with the Integrating Group, are producing reports which will not only be relevant to the Programme as a whole, but will also stand on their own feet as authoritative documents within the area of specialism of the authors. This fact represents an important added incentive to participation.

We feel very strongly that any transdisciplinary ecological study of a human settlement must, if it is to be successful, have a sound conceptual basis. Research in human ecology is not just a question of measuring as many variables as possible, feeding them into a computer, and expecting something useful to come out. Considerable effort has therefore been spent on the development of the conceptual framework of the Hong Kong Programme. Time does not permit me to discuss this in any detail here, but I will attempt to summarise the main points. On the one hand we are interested in certain important ecological characteristics of the system as a whole, in particular the patterns of flow and utilisation of energy and of certain important materials, such as water. This kind of information, apart from its obvious relevance to various problems associated with the supply of energy and important material resources, contributes to an understanding of important dynamic interrelationships and interdependencies within the system. Colloquially speaking, it helps us to comprehend how the whole thing works. As I shall discuss in a moment, it is also relevant to the health of the human population. The pattern of energy flow will be described in terms of its spatial, institutional and socio-economic distribution, and also in terms of the ultimate effects of its uses on human experience.

Another factor influencing the conceptual approach is

our interest in the important interrelationships between environmental variables and general conditions of life on the one hand and human health and well-being on the other. It is well accepted that health is affected by the quality of the environment and by life style, but we need to know a great deal more about the nature of these effects. It is important to bear in mind that there are, as far as the individual's response is concerned, two levels of "environment" (see Figure 2). There is what we refer to as the "total environment" which includes all the variables which comprise the system as a whole. And there is the individual's personal environment, which is actually experienced by the individual. It is, of course, the personal environment which directly affects the individual's biopsychic state - that is, his actual physical and mental state, and hence his health. The personal environment is a function of the properties of the total environment and various "filters" which separate the individual from the total environment.

Thus, we can recognise three levels of variables relevant to the human state in a settlement. The first level comprises the biopsychic state variables, relating, for example, to health and disease, growth rates and fertility and mortality patterns. The second level - the level of personal environment - comprises the actual conditions of life of individuals and includes, for example, such material variables as diet, quality of air inhaled, size of dwelling, population density in dwelling, noise levels, and such psycho-social and behavioural variables as degree of co-operative small group interaction, degree of personal creative behaviour and degree of emotional involvement and sense of challenge in daily activities. At the third level - the level of the total environment - we recognise four categories of variables as follows:-

1. The non-biotic components of the environment.
2. The biotic components (other than human).
3. The human population as a whole and its social institutions.
4. The cultural components (value systems, beliefs, laws, etc.).

We have found this model to be very useful in the Hong Kong study, since it provides a rational means of organising variables and of describing principles and hypotheses concerning interactions between variables within different categories. Unfortunately, it is not possible in the time available to discuss the various hypotheses which are being tested in the Programme, and it must suffice to say that the topics covered range from the effects of "crowding" (personal environment) on health (biopsychic state) to the interrelationships between the pattern of energy flow in the total environment and patterns of health and disease in the community.

Before leaving the conceptual aspects of the Programme, I would like to emphasize that quantifiability has not been one of our criteria for selection of variables for consideration. In fact, we consider that it is unscientific to ignore variables which, although judged to be relevant, are not easy to quantify. There is no law of nature or of society that states that there is likely to be any relationship whatsoever between quantifiability and importance. Thus, in the case of variables which are difficult to quantify but which we consider important, we simply measure them as best we can. If necessary, we are not afraid to resort to an impressionistic approach, although we must always, of course, bear in mind the pitfalls inherent in the interpretation of information derived in this way.

The present situation with respect to the activities of the Integrating Group in the Programme may be summarised as follows. Our year of data collection, 1974, was highly successful, and our main difficulty now lies in the fact that we have acquired vastly more information than we had anticipated. The Biosocial Survey was completed in 1974, and the results which are just now becoming available for analysis promise to be very interesting. Much of the work on energy flow and utilisation is completed and three papers on this aspect, as well as one on our theoretical approach to the problem of crowding, are already in the hands of publishers. However, the main work in our "phase of integration and synthesis" still lies ahead, and we hope to have completed it by this time next year.

The studies have been supported financially by the Australian National University and by a grant from the Nuffield Foundation. Very recently the Programme has been formally adopted by UNESCO as a pilot study for its MAB Project 11, and in this connection some further funds have been made available by the United Nations Environmental

Programme.

Before concluding, I would like to stress one further point which we consider a very important feature of the Hong Kong Human Ecology Programme. At no time since the very beginning of the Programme did we suggest that we were setting out to advise or make recommendations to decision makers, planners or government departments with respect to their environmental policies. We have seen our task as an attempt to improve understanding of the total ecological situation in Hong Kong (and hence, also in other settlements), and then to communicate our findings to anyone who may be interested, including members of Government Departments. Needless to say, we hope that work of this kind will, through the improved understanding that it brings, contribute to the attempts of human society to plan wisely for the future.

In retrospect, we see many serious deficiencies in the Hong Kong study. It was a Programme which just "evolved" and it lacked the advantages of thorough planning in advance. For example, there has been little synchronisation and coordination in the activities of the various Specialist Groups. In any similar Programmes in which we might be involved in the future, we will attempt to ensure that the Specialist Groups commence their projects at about the same time and that they cover the various aspects of the ecology of the settlement in a somewhat more systematic way. Another serious weakness of the present study has been the shortage of senior staff in the Integrating Group. The full responsibility for much of the Programme in Hong Kong was, for long periods, carried out by the two Ph.D students, Sheelagh Millar and Ken Newcombe, due to the fact that I, the Director of the Programme, was forced to be away in Australia because of other commitments. While the students performed this task admirably, it was unfair to burden them with this responsibility. Future programmes of this sort should have a permanent professional staff in the Integrating Group of at least three persons.

In concluding, I would like to say that for those of us actually involved, the Hong Kong Programme has been a unique and exceptionally rewarding experience, both intellectually and emotionally. Needless to say, we are personally very appreciative of the part played by CHEC in bringing this about.

SUGGESTIONS FOR A CONCEPTUAL BASIS OF A PROGRAMME OF INTERNATIONAL ECOLOGICAL STUDIES ON HUMAN SETTLEMENTS

STEPHEN BOYDEN

Aims and Objectives

Broadly, the aims of the international comparative ecological studies on human settlements should be as follows:-

1. To improve knowledge and understanding of the ecology of each human settlement as a whole, in terms especially of the patterns of flow and utilisation of energy and of certain important materials, and taking into account the interrelationships between the settlement and its surroundings.
2. To improve knowledge and understanding of the causal interrelationships in human settlements between a) the properties of the environment, b) the human conditions of life and c) the health and well-being of the population.
3. To improve knowledge and understanding of the interrelationships between the ecological properties of the human settlement as a whole (e.g. patterns of energy flow and utilisation) and the health and well-being of the people.
4. To improve knowledge and understanding of the cultural processes of adaptation to detrimental environmental influences.

The organisation of the programme would be based on the overriding view that such improved understanding of human situations requires an integrative approach which brings together findings and concepts from the natural sciences, social sciences and humanities as they relate to these situations. It is believed that through the improvement in understanding that they bring about, integrative studies on human settlements of the kind proposed will contribute to the development of wise social policies aimed at maintaining and improving levels of human well-being and at safeguarding the life-supporting systems of the biosphere on which mankind depends.

The general approach should be 1) to integrate knowledge already acquired in different areas of specialism by academic, governmental and other authorities which can contribute to the comprehensive understanding of the situation and 2) to initiate such new investigations as are deemed necessary and feasible to further such understanding.

For integrative studies of this kind to be successful and worthwhile, it is essential that each be based on a clear and definite conceptual framework, the theories, postulates and hypotheses of which provide the studies with direction and structure and determine the selection of variables that are taken into account.

The present paper proposes a model for the conceptual framework of the programme.

General Theoretical Considerations

The starting point for the construction of the proposed conceptual framework is provided by biological science. There is a certain logic in this, since in evolution biotic processes preceded and gave rise to cultural processes, since they are a necessary basis for the perpetuation of cultural processes, and since every human organism begins life as an exclusively biotic being. Nevertheless, it is appreciated that the human situations which are the subject of the proposed studies involve a continual dynamic and highly significant interplay between natural (biotic and non-biotic) processes and cultural processes (1), and that the proper ecological study of these situations must therefore necessarily take account of variables normally studied by students in a wide range of academic disciplines. The value systems of a society, for instance, are as important as determinants of its ecological state as are the available sources of fossil fuels and foodstuffs.

Human Ecology

The overall design of the proposed studies would be

based on the two conventional approaches to the study of ecology in biological science, namely, system ecology and population ecology (2). In system ecology, ecosystems as a whole are the object of study, and especially the patterns of flow of energy and of important materials or substances within the system. In population ecology emphasis is on a given population and the important interrelationships between this population and the non-biotic and the other biotic components of the total environment (3).

The conceptual basis of these two aspects of the programme and the approach to the study of the relationships between them is summarised below.

System Ecology

Ecosystems which include human societies contain three sets of interacting variables; the non-biotic components, biotic components, and the cultural components. An urban ecosystem as a whole is depicted in the model in Figure 1. Because in most city ecosystems by far the most important biotic population, in terms of biomass and of influence on the system, is that of mankind, the human population is represented in the diagram separately from the other biotic components of the system.

A human settlement is, of course, a dynamic system and its integrity requires an input and outflow of energy, and an organised circulation within and through the system of numerous materials. Knowledge of the patterns of energy flow and of the patterns of flow of important substances is essential for the proper understanding of the multiple subtle interrelationships in the system.

The dynamism of any ecosystem is the consequence of processes taking place within the system. Thus, in describing the place of a population in a natural ecosystem, one is concerned not only with its numbers, distribution and biomass, but also with what the population and its members do. In the case of animals this means both behaviour as it relates to ecology (e.g. hunting, foraging, migration, etc.), and environmentally relevant aspects of metabolism (e.g. consumption of O_2 and release of CO_2). Important processes involving plants include photosynthesis and growth.

A processes dimension also exists with respect to the non-biotic set of components, although in the case of natural ecosystems these are limited to such processes as, for example, the evaporation of water or its flow along the rivers, and changes consequent upon movement of the earth in relation to the sun, or the moon in relation to the earth. In human settlements, however, the processes dimension of the non-biotic components of the system has acquired additional importance, and involves the activities of machines of many different kinds. In fact, by far the greater part of the energy flowing through modern urban settlements is used, not in the activities of human beings or other biotic populations, but in the activities of machines.

While the biotic and non-biotic components of the system clearly have processes dimensions which involve the use of energy and affect other components of the system, it is more difficult to conceptualize the processes dimension of the cultural components of the system. Nevertheless, it is plain that beliefs, ideas and laws, for example, are active in the system, but only through the processes dimension of the human population; that is, through human behaviour.

In fact, of course, culture has no impact on the system nor do any components of the system have impact on culture, except through the agency of the human population. These facts draw attention to an inaccuracy in Figure 1, in that arrows of influence pass directly in both directions between culture and non-biotic components and between culture and biotic components. However, for the sake of simplicity, this convention will be retained in this paper, and the involvement of human behaviour in these interactions will be taken for granted.

A study of the inputs and distribution of energy, and the uses to which it is put, represents a logical and useful entry point to the description of an ecosystem or of a human settlement in terms of system ecology. Although a considerable amount of work has been carried out on the energy budgets of natural and agricultural ecosystems, very few studies have so far been made on the energetics of human communities. This is a serious deficiency in the present era of impending energy crises, and there is an urgent need for more comparative data on the patterns of energy flow in urban and other human situations. Only on the basis of such data will it be possible for society to develop intelligent policies for coping with energy problems.

One of the major premises influencing the proposed conceptual basis of the programme is the view that any satisfactory and lasting solution to the energy crisis will involve not only technological innovations and the exploitation of new sources of energy, but also, and probably more importantly, a degree of reorganisation in society (e.g. involving commuting patterns, transport systems, organisation and kind of industry, distribution of work-force). Consequently, in the proposed studies the energy flow should be analysed and described in terms of 1) spatial distribution; 2) the institutional and collective uses to which it is put (e.g. public and private transport, labour-intensive and non-labour-intensive industry, social services, waste disposal) and 3) individual human experience and behaviour (e.g. recreation, geographical mobility, heating, cooling, cooking).

An important challenge in the ecological study of human settlements systems is to attempt to recognise significant interrelationships between the findings at the level of the system and those at the population level. It may be assumed, for example, that important relationships exist between the pattern of energy utilisation in a settlement and the state of health and well-being of the population, and that changes in the former will have important repercussions on the latter. However, although bold assumptions are commonly made on the nature of these relationships, very little is known about them.

As already mentioned, system ecology also includes analysis and description of the patterns of flow and distribution of certain materials in the system, and an important, but complex relationship obviously exists between the patterns of energy flow and the patterns of materials flow. Indeed, all movement of material involves, or is dependent on, a flow of energy. One of the most important materials in an urban system is water. It is suggested, therefore, that each of the studies should include a detailed analysis of the pattern of inputs, distribution and outputs of water. Flow charts should be constructed of the supply and uses of water in the region and consideration should be given to the role of water as a carrier of biocides, fertilizers, viruses, human and animal sewage and general waste material.

Some categories of variables in the total environment that are likely to be considered important are listed at the end of this paper.

Population Ecology

In population ecology, the investigator concentrates his interest on a population of a certain species (in the present context, the human species) and on its important interrelationships with biotic and non-biotic components of the system. The population ecology approach is applicable to the whole population, to sub-populations and even to individuals.

For the purposes of the present discussion, it is useful to recognise three levels of interdependent variables in the ecological study of human populations. We shall refer to them under the following headings:- 1) biopsychic state, 2) life conditions, 3) total environment and filters. The total environment is, in essence, the system as a whole, as described above under "system ecology". In describing these three levels of variables, it is convenient to put the emphasis on individuals, although the principles also apply, with slight modification, to sub-populations and whole populations.

Level 1 - The Biopsychic State. The Biopsychic state includes all those variables which describe the biotic and psychic characteristics of the individual (or of the population). It includes anthropometric variables (e.g. height, adiposity) and variables relating to genetic characteristics, nutritional state and general physical and mental state (e.g. blood pressure, presence of infection, measures of mood, personality and mental health). The values, attitudes and knowledge of an individual are also an aspect of his biopsychic state. In the case of populations and sub-populations, the biopsychic state includes rates of fertility, mortality and incidence and prevalence of disease.

Level 2 - Life Conditions. The biopsychic state of an individual is influenced to a large degree by the actual conditions of life experienced by the individual (which, as we shall see, are related to, but not a direct function of, the total environment). We recognise two aspects of life conditions, which are referred to as a) Personal environment and b) Personal behaviour.

a) Personal Environment. The personal environment includes all the variables which describe the individual's personal experience of the total environment (i.e. of the system as a whole). They thus include the four sets of variables mentioned above in relation to system ecology, that is, non-biotic components, biotic components, human population (i.e. the rest of the human population) and culture. The categories of variables likely to be taken into account in integrative ecological studies on human populations are listed at the end of this paper. They include experience not only of material variables such as food quality, quality of air inhaled, but also psycho-social variables such as experience of small-group interaction and of sense of challenge offered in daily activities.

b) Personal Behaviour. The biopsychic state of an individual is also affected by what the individual actually does in his personal environment (e.g. physical work, smoking habits). The personal behaviour variables are, of course, influenced by, and they also influence, the personal environment. They also influence and are influenced by the biopsychic state.

It must be stressed that the classification of variables which is suggested here, while conceptually and practically useful, is not intended to be a rigid one; nor is it intended to imply that the divisions between the categories are sharp and clear-cut. As is so often the case in biosocial science, sharp divisions between the different related categories of variables are neither possible nor meaningful. For instance, the degree of small-group interaction experienced by an individual is a function of both his personal environment (the behaviour of other people in his environment) and of his personal behaviour (his own behaviour in that environment). Similarly, his personal behaviour, which we have chosen to link, under "life conditions", with his personal environment, could also reasonably be considered as an aspect of his biopsychic state.

Level 3 - The Total Environment. Clearly, the personal environment of an individual is related to the characteristics of the total environment. The relationship is not, however, a direct one. For example, there may be a wide range of foodstuffs available in the total environment, but only a few of these may be actually consumed by the individual. In other words, certain factors, such as an individual's economic status or certain cultural taboos, restrict his experience of the total environment. These factors which determine the relationship between the individual's personal environment and the total environment are referred to collectively as filters. The most important of these relate to social values, social norms, and economic status.

As mentioned above, the total environment is essentially the ecosystem as a whole, and it includes the four sets of variables referred to under "system ecology" above. For practical purposes, even when the whole population is the subject of the population ecology approach, it is useful to retain the "human population" set of variables in the total environment. Thus, such categories of variables as societal organisation and social services are most effectively considered as part of the total environment rather than as an aspect of life conditions.

The interrelationships between these sets of variables are illustrated diagrammatically in Figure 2.

The biopsychic state of an individual at any given time is determined not only by his life conditions at that time but also by his genetic constitution or genotype and by his previous life conditions, from the moment of conception onwards. These influences are included in the model in Figure 3. Another factor (also depicted in Figure 3) which plays an important part in some instances in determining the response of an individual to his life conditions is his perception of components of his personal environment. Thus, the individual's biotic response to noise can be greatly influenced by whether or not he perceives the noise as a threat. Perception itself is a function of previous interactions between personal environment and genotype.

Evolutionary Considerations

The interrelationships and interdependencies in a

natural ecosystem are, of course, products of the processes of biotic evolution, as are the genetic characteristics of the component biotic populations in the system. The principles of biotic evolution are therefore very pertinent to understanding ecosystems in general, including human settlements.

In human population ecology we are especially interested in the interrelationships between variables in the total environment, variables in life conditions and variables in health and disease. In other words, we are interested in the consequences for human health and well-being of variations and changes both in the total environment and in the life conditions. This includes a general interest in the ways by which cultural developments or cultural differences may, through their influence on the biotic and non-biotic components of the environment and on life conditions, affect the biopsychic state of individuals. A further area of interest centres on the patterns by which human populations respond to, and attempt to adapt to undesirable environmentally-induced changes in the biopsychic state.

One evolutionary principle is of special relevance to the understanding of the responses of human beings to their life conditions. It is referred to as the principle of phylogenetic maladjustment. This principle is a corollary to the Darwinian theory of evolution, according to which species become, through natural selection and over many generations, increasingly well adapted in their genetically-determined characteristics to the conditions of life in the environment in which they are living. It follows from this fact that, if the environmental conditions deviate significantly from those prevailing in the evolutionary environment, it is likely that individuals will be less well-suited in their biotic characteristics to the new conditions, and consequently some physiological or behavioural signs of maladjustment may be anticipated. Maladjustments due to the exposure of animals to conditions which deviate from those of the evolutionary environment are termed "phylogenetic" because they reflect the evolutionary experience of the species and are, in the changed environmental conditions, characteristic responses of the species as a whole.

Again, it is necessary to emphasize that, even in the

case of phylogenetically-determined reactions to evodeviations (4), there are likely to be variations among individuals in their sensitivity to given environmental pressures. These variations may be due to differences in previous life conditions or in individual genetic characters. It should be noted that, while evodeviations are likely to give rise to phylogenetic maladjustment, they do not necessarily do so. For example, the removal of an important predator from the environment will not give rise to phylogenetic maladjustment, although it may well result in a disturbance in the population dynamics of the population affected.

In nature, if an evodeviation which causes phylogenetic maladjustment persists, the affected population is likely either to become extinct or, eventually, to become genetically adapted to the new conditions through natural selection.

One of the outstanding biological effects of civilisation during the last ten thousand years has been the extent to which it has brought about changes in the biotic conditions of life of the human species. As would be anticipated, since the principle of phylogenetic maladjustment applies as much to Homo sapiens as it does to any other species, these evodeviations have led to numerous examples of phylogenetic maladjustment in response to these culturally-induced environmental changes. The important differences between the human species and other animals is that mankind has the advantage of an extra set of adaptive processes, the processes of cultural adaptation, which can render individuals and populations better able to cope with the new conditions.

The significance of the principle of phylogenetic maladjustment for biosocial studies lies in the fact that it provides guidelines for the selection of life condition variables for measurement, and is a fruitful source of hypotheses relating to the influence of environmental conditions on the biopsychic state. We know, of course, that mankind evolved as an omnivorous hunter-gatherer, and that the phylogenetic characteristics of the species were determined by the selection pressures operating in that primeval environment. Tens of thousands of generations of hunter-gatherers were to pass before even the first cities came into existence only 200 generations ago. Bearing in mind also that only a very small minority of the human population has, in the intervening period, actually lived in cities, we can safely say that natural selection has not produced a new breed of Homo sapiens better adapted in its genetic characteristics than pre-urban populations to the conditions of city life. While we must be clear in our minds that one of the important features of the domestic transition was the protection it afforded human beings from many of the hazards and causes of death in the primeval environment, we must also appreciate that people were at the same time exposed to a series of novel life conditions. Each of these represents an evodeviation, and is therefore to be considered a potential source of phylogenetic maladjustment. It follows that knowledge of the conditions of life and the behaviour of primeval or hunter-gatherer people provides a biologically valid starting point for the selection of life condition variables for measurement in human population ecology studies, the general hypothesis being that all recognisable deviations from these conditions are under suspicion as causes of ill-health, until proven otherwise.

The principle of phylogenetic maladjustment applies without question to the "material" aspects of life conditions such as food quality, air quality, contact with noxious chemicals etc. About this, there can be no debate. An important postulate or broad hypothesis which is relevant to the proposed studies is that the principle applies also to what we might term "psychosocial" aspects of life experience. Indeed, in some areas this hypothesis is certainly true. Enforced solitude for long periods, for example, is well known to give rise to signs of maladjustment. However, we suggest that this principle may well apply to many other aspects of life experience of the individual, such as the degree of small group interaction, the level of personal creative activity, and the sense of challenge in daily experience. The phylogenetic maladjustment resulting from either material or psychosocial evodeviations may be physical (e.g. atherosclerosis) or psychological (e.g. psychoneurosis) in nature.

Adaptation

It is not necessary to stress that human settlements are dynamic and constantly changing systems. Thus, a new cultural development may bring about a change in the life conditions of a section of the population either directly (e.g. by affecting the psychosocial conditions of life of individuals) or indirectly by first affecting the quality of the biotic or non-biotic environment. If the change in life conditions represents an evodeviation, some signs of maladjustment are likely to occur, and if these are sufficiently serious to cause concern, society may respond by introducing measures aimed at overcoming the maladjustment in some way. This brings us to the fourth objective of studies in urban population ecology (see page 1) - the improvement of understanding of the various processes by which societies adapt, appear to adapt or attempt to adapt to signs of maladjustment in the biopsychic state of the population resulting from unsuitable life conditions. For example, through cultural processes, steps may be taken to reverse a change in the material environment that has produced an evodeviation in life conditions and consequent signs of maladjustment. This sequence of events is represented diagrammatically in Figure 4.

We can recognise several different classes of cultural adaptation to phylogenetic maladjustment, which differ with respect to the rationale behind them, their mode of action and their long-term effectiveness. For example, in some cases the aim of the cultural response is to reverse the underlying change responsible for the state of phylogenetic maladjustment. This is the form represented in Figure 4 and is known as corrective cultural adaptation. In contrast, in antidotal cultural adaptation, the attempt to overcome the disorder is directed at the symptoms of the disorder or at an intermediate cause, but not at the unsatisfactory biological conditions which gave rise to the state of maladjustment in the first place. Frequently antidotal cultural adaptation not only leaves unchanged the initial evodeviation, but also introduces a new one. Experience has shown that, in general, corrective adaptation is usually more successful than antidotal adaptation in the long run, although it is not always a feasible proposition for a variety of reasons.

Description and the Testing of Hypotheses

It is suggested that the selection of variables for study in integrative studies of human settlements be based on two considerations. In the first place, the studies should aim simply to describe the settlements in terms of obviously important ecological parameters (e.g. patterns of energy flow, population size, population density, mortality and fertility rates, air quality, prevalence of diseases, nutritional data, etc.), and in terms of special aspects of the situation which the conceptual framework or the local conditions suggest are important. Secondly, because the conceptual framework brings to mind a number of hypotheses concerning the nature of the interrelationships between components of the system (e.g. cultural components, personal environment and biopsychic state), integrative ecological studies provide an opportunity for testing some of these hypotheses.

Space does not allow further consideration here of the nature of the hypotheses that might usefully be examined in a cross-cultural programme of ecological studies on human settlements. However, it is strongly recommended that the organisation of such a programme should allow the free exchange among participants in different localities of ideas concerning relevant hypotheses that might be tested.

CATEGORIES OF VARIABLES

It must be stressed again that this categorisation of variables is not in any way absolute, and it must also be emphasised that the theme of the integrative approach is the interrelationships between variables in different categories. The studies should not, therefore, take the form of a simple listing of data, according to such a categorisation, of the selected variables. The categorisation is merely an aid to the selection of variables for measurement and to the formation of hypotheses concerning the interrelationships between variables of various kinds.

The Total Environment (5)

Non-biotic components and processes.

General topography

Climate

Somatic energy -

solar irradiation - total input and contribution to somatic energy budget; imported somatic energy;

distribution - spatial, socio-economic, etc.
Nutrients (e.g. protein, fats, minerals) - inputs; distribution (spatial, socio-economic etc.)
Pharmaceutical products and psychotropic drugs
Chemical contaminants of foodstuffs
Organic wastes
Extra-somatic energy - inputs; distribution (spatial, socio-economic); end use in terms of activities of society and social organisation (e.g. industry, transport, commerce, construction, commuting); end use in terms of individual human experience (e.g. heating, lighting, cooking, recreation); outputs (in terms of material exports); incidental aspects (e.g. climate modification).
Water - patterns of flow and utilisation
Other selected materials - patterns of flow and distribution
Inorganic wastes
Gaseous exchange and air quality
Buildings, streets, open places
Noise levels
Transport systems

Biotic components
Animals
Plants
Parasites and pathogens

Human Population
Biotic variables - e.g. population dynamics; mortality and fertility rates; age structure; population distribution; population densities; population movements; nutritional state; growth and maturation rates; genetic characteristics; ethnic characteristics; physical fitness; prevalence and incidence of physical and mental disorders
Societal variables - e.g. social organisation; socio-economic structure; family structure; other groupings; social services.
Behavioural variables - e.g. group behaviour; institutional or corporate behaviour; crime rates; commuting behaviour; occupational behaviour

<u>Culture</u>
Historical and cultural background of population
Value system (e.g. religions; attitudes towards technology, material wealth)
Aspirations
Legislation
Knowledge
Technology

Life Conditions

<u>Personal Environment</u>
Non-biotic -
e.g. air quality; food quality and quantity (calories and nutrients); exposure to noxious chemicals; ionic irradiation; potential sources of severe physical injury; noise levels; structure and size of dwellings; use of pharmaceutical products and psychotropic substances
Biotic -
e.g. exposure to potentially pathogenic micro-organisms and to metazoal parasites; interactions with animals
Human population -
e.g. population density; number of daily social interactions; family structure; in-group/out-group structure; support networks; small group interaction patterns; experience of criminal or violent behaviour
Culture -
e.g. values and social norms directly influencing individual(s); legislation; learning experience; aspirations of in-group(s)
Other -
e.g. degree and variety offered in daily activities; degree of challenge and sense of purpose offered in daily activities; opportunities and incentives for exercise of creative activities and the exercise of learned manual skills; degree of emotional involvement experienced in daily activities

<u>Personal Behaviour</u>
Elemental behaviour -

e.g. feeding and drinking behaviour; physical work and activity; resting and sleeping patterns; sexual behaviour; social interaction patterns; aggressive behaviour; play and active entertainment; passive entertainment; creative behaviour and the exercise of learned manual skills; learning behaviour

Biological time budgets

Biopsychic State

Anthropometric variables -
e.g. height, weight, skin folds

Physiological variables -
e.g. blood pressure; lung capacity; body temperature; measures of physical fitness

Psycho-social variables -
e.g. performance in various psychological tests

General level of health -
e.g. as indicated in physical and mental health scales

Specific signs of physiological or psychological maladjustment

Genetic variables -
e.g. as indicated in various tests for genetic markers

(Note: the individual's personal knowledge, values and attitudes are also aspects of his biotic state)

Filters

Socio-economic, cultural and other factors considered important as determinants of the individual's experience of the total environment.

Perception

Variables which can be used as indicators of the individual's perception of certain aspects of his or her personal environment (e.g. expressed attitudes towards noise levels, crowding, etc.)

(Note: It will be noted that some of the variables under "total environment" also appear under "life conditions" and "biopsychic state" headings. This fact does not reflect any inconsistency in the conceptual model. In any particular study, or any particular part of a study, the classification of such variables depends on the nature of the questions which are being asked. For example, one may be interested in the values held by members of a population (or sub-population) as influences on their behaviour patterns. In this case - they would be considered an aspect of the "biopsychic state". Alternatively, one might be interested in how the values of the society affect the behaviour of members of a population (or sub-population), regardless of whether the individuals share these values. In this case, the values belong in the "culture" component of the total environment. Similarly, violent behaviour may be of importance to individuals either as an aspect of their own "personal behaviour" or as a feature of the behaviour likely to be encountered in the "total environment" (and hence in the "personal environment").)

NOTES

(1) For the purpose of this paper, the term "natural processes" means processes of a kind that existed on earth before the advent of mankind. By "cultural processes" we mean the special processes which are characteristic of human societies and which depend on the acquisition and accumulation of information and its transmission by non-genetic means, mainly through the use of learned symbols, from one human being to another, from one society to another, and from generation to generation.

(2) While this separation - system ecology and population ecology - is useful both conceptually and practically, it is nevertheless an oversimplification of the situation, and in human ecology especially it is necessary to recognise other levels of interest and concern, ranging from the biosphere as a whole, through ecosystems, populations and sub-populations to the individual.

(3) In this paper the term "total environment" is used to denote all components of the ecosystem of which the population of interest is a part.

(4) Deviations from the conditions of life that prevailed in the evolutionary environment of a species are henceforth referred to as "evodeviations".

(5) In this list, no attempt is made to separate the static components of the system from the processes, although the conceptual appreciation of the distinction between these two aspects is essential for the proper understanding of the study.

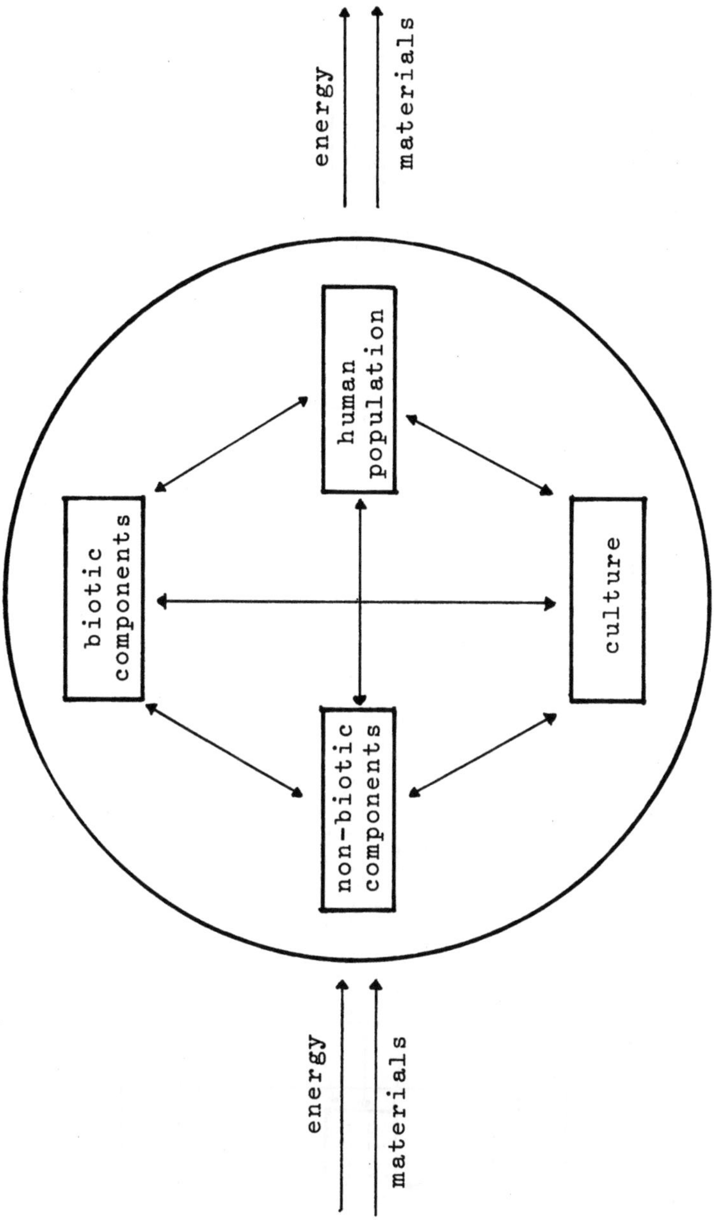

Fig. 1 System ecology of a system

Fig. 2 Population ecology of *Homo sapiens* (simple version)

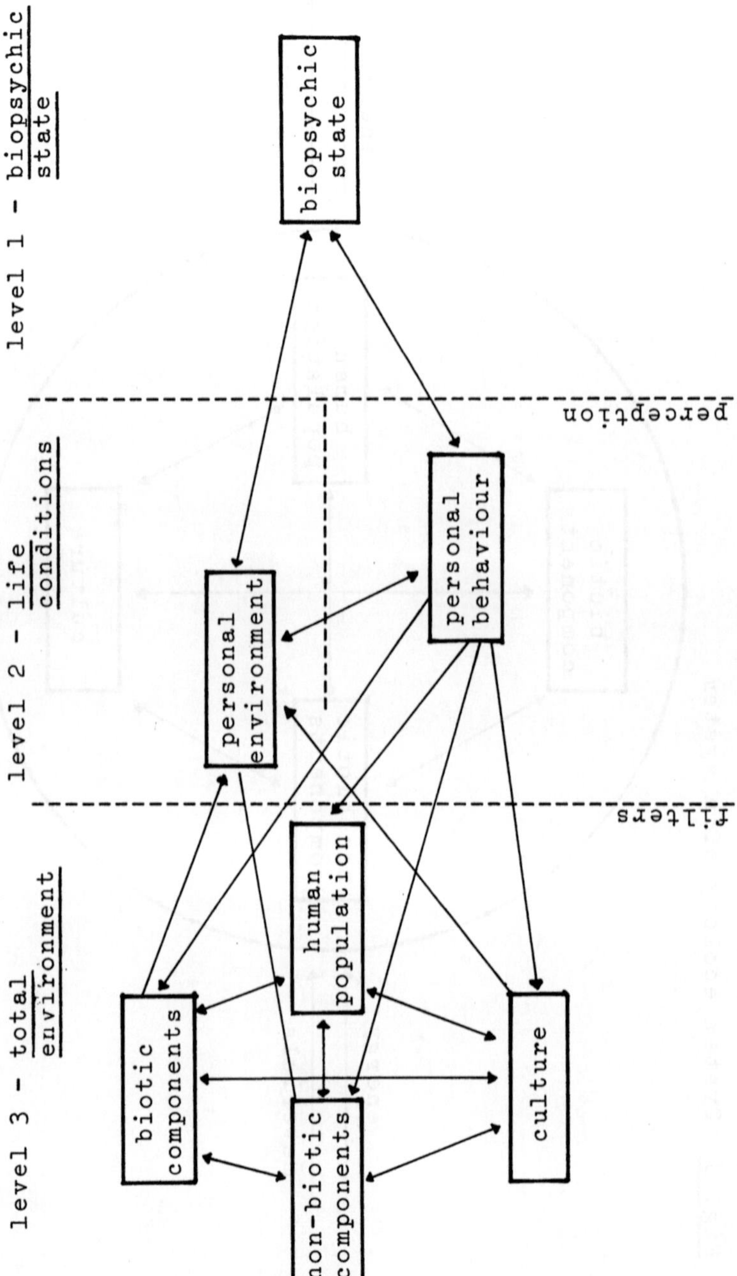

Fig. 3 Population ecology of *Homo sapiens* (taking previous experience into account)

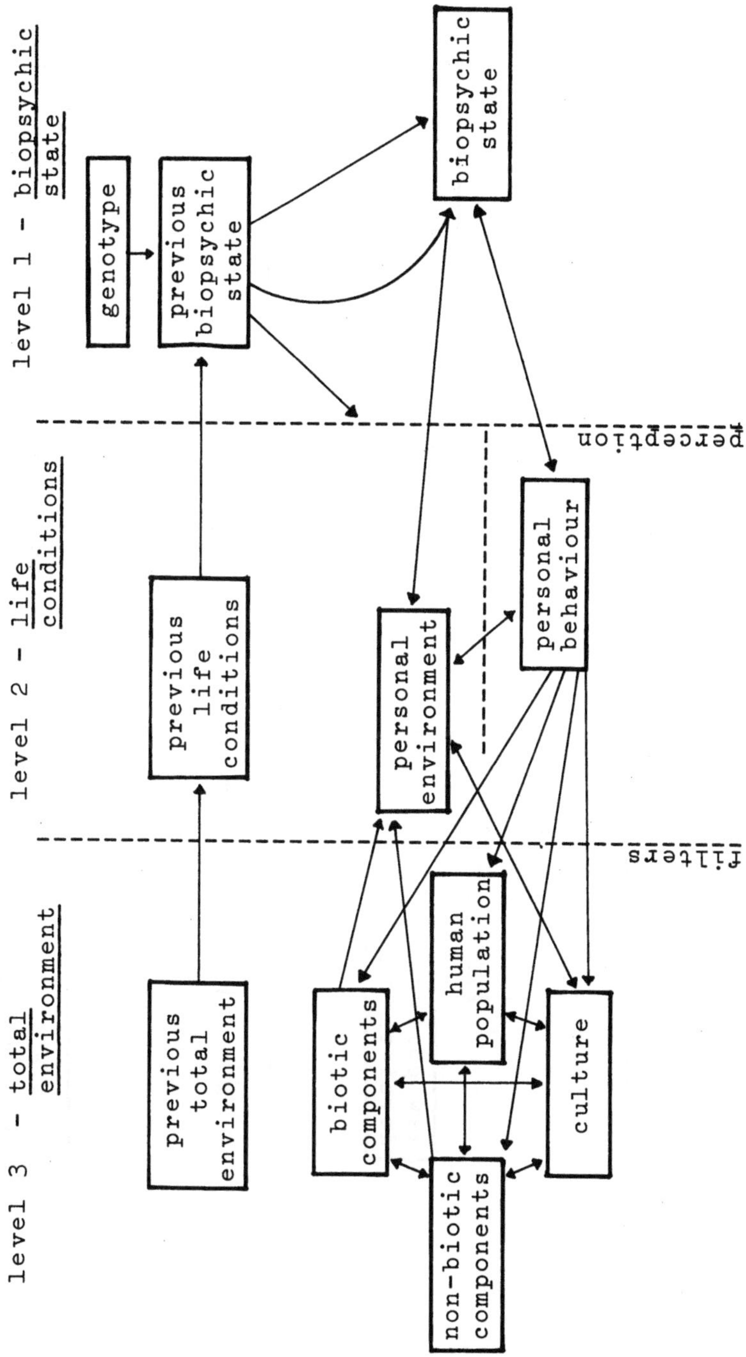

Fig. 4 Corrective cultural adaptation to phylogenetic maladjustment

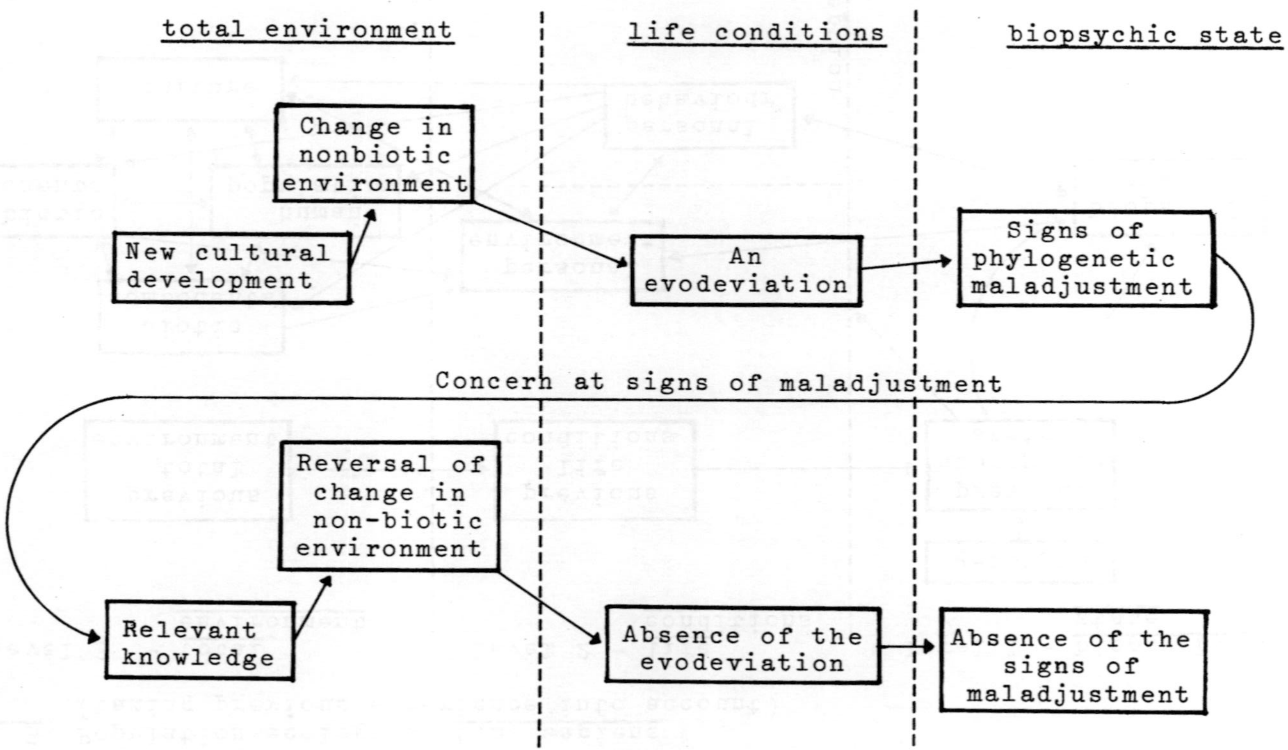

HUMAN ECOLOGY AND PLANNING FOR HUMAN SETTLEMENTS

PERCY JOHNSON-MARSHALL

Introduction

In this last lecture in the series it may be appropriate to commence by recording briefly some aspects of the themes developed by the other contributors. The general aim of the programme was to consider problems of human settlements from different points of view, taking advantage of the variety of disciplines represented in CHEC, and of the wide range of Commonwealth experience.

As a result a valuable contribution was made to certain fundamental problems facing human society in countries of widely differing economic levels. At one extreme Messrs Gertler and Danson dealt with problems of urban concentration and population movements in a country with one of the highest standards of living in the world. At the other extreme Messrs Kutena and Obeng dealt with the basic problems of low income countries. Professor Kutena emphasized that in the case of Bangladesh it was a matter of human survival, and that an incorrect decision could endanger many human lives. Both Messrs Obeng and Isaacs stressed the need for water as a basic resource for human livelihood, and Dr Obeng emphasized how a large scale project linked to a water resource could influence human resettlement. Professor Isaacs emphasized the role of government in Jamaica in responding to the need for environmental protection, while Professor Danson related governmental action in Canada to problems of population imbalance. Professor Gertler described the dilemma of human settlements in a country such as Canada, which is heavily influenced by the private sector, while Professor Roberts stated some of the problems and interests of large and small societies in New Zealand, with emphasis on the need for smaller settlements to be met in a human way. Dr Boyden brought out some fundamental aspects of the complexity of problems of ecology and human settlements and developed the theme of the necessity of a multi-disciplinary approach.

This multi-disciplinary concept is, of course, at the foundation of CHEC's role as a Commonwealth organization, since it brings together the widest range of disciplines in order to consider the widest range of human interactions with the environment.

The series has emphasized a new and important trend in human society, where each specialist understands more clearly the interrelationship of one expert contribution to that of all the others; and becomes aware that, no matter how important his or her own activities are, they may even be harmful if they are not linked across the board within a broad ecological framework.

It is almost a truism to say that post industrial society has suffered too long from uneven scientific progress, where the most dramatic advances in a particular sector have caused acute imbalance in others. Too many outstanding examples of this narrow and obsolete specialist approach still exist in many parts of the world, and one has only to consider the opening up of large areas of so-called virgin land for farming in order to produce important new crop yields, to find in a few years that such large scale operations, while having short term advantages, have in some cases had serious long term defects in terms of soil erosion, loss of soil fertility, etc. The lessons of such classic cases as the Texas Panhandle or the African Ground Nuts Scheme have been slow to be learned, and this series has, above all, enlightened the mutual responsibility of specialists.

To some extent Professor Gertler and I have been fortunate in our applied discipline of planning, in that the planning profession consists essentially of experts from different fields applying their knowledge to the planning of the environment, and for many years it has had a built-in inter-disciplinary approach, which is now being seen to have wider implications for ecology as a whole.

The United Nations Context

The origin and purpose of the Commonwealth Human Ecology Council has been closely linked to the United Nations Environment Programme, and more particularly to the famous Stockholm Conference on the Environment which originated the Programme itself. It was significant that the governing council of the Programme decided at the earliest possible time to organize a United Nations conference on Human Settlements in Vancouver, and to stress the important contribution which could be made by non-government organizations (N.G.O.s) at the conference. In fact, the progress report (1) of the governing council presented at Nairobi in April of this year states that "the active participation of non-governmental organizations is essential in order to achieve the proper impact and follow through on which the success of the conference depends" and that it was envisaged that a parallel conference Forum would be organized at the time of the main conference.

CHEC has, through its Secretary General, been in close touch with the N.G.O.s working group, and its policy in setting up a special Habitat committee to prepare a report for the Forum is indicative of the importance which is attached to the whole concept and significance of the conference.

As the United Nation's consultant on Human Settlements for the Stockholm Conference, I have been particularly happy to note that the majority of the recommendations contained in the Position Paper (2) which I was required to submit to the Conference Secretariat, were adopted and are included, in some cases almost to the identical words, in the Conceptual Framework. This, of course, need cause little surprise, since it was based on many earlier U.N. documents, and more particularly on those published by the U.N. Centre for Housing, Building and Planning (3).

Settlements and Human Ecology

The Position Paper which I have mentioned pointed out that the problems of human settlements were at the root of many terrestrial ecological problems today, and that a unique opportunity existed among the nations of the world to collaborate in sharing experiences of their problems, their knowledge, ideas and resources, in order to aim at the achievement of a high standard of civilization accompanied by a high quality of environment. It was pointed out that one of the ways by which these objectives could best be achieved was by the good planning and management of human settlements, wherever they might be.

The Commonwealth as an organization offers an excellent prototype for the consideration of such problems, since it contains within it almost all kinds of settlement and all economic levels. Geographically, it covers land problems from the world's largest states down to the intricate space problems of small islands; and climatically it includes polar regions on the one hand and the hottest tropical zones on the other. Its human settlements vary in size from the smallest and most primitive villages to some of the largest and most sophisticated urban agglomerations in the world. Within its collective experience, therefore, one would expect to find every kind of problem being faced, if not being resolved.

As a planner I would like to put forward some of the contributions which environmental planning can make in the wider ecological context. As I see it every expert and every specialist has certain common objectives. Fundamentally, all our work is concerned with human welfare and improving human life on the planet. The planning of settlements is only one aspect of the broader problems of social welfare and could be said to fit into the general pattern of social services. It is concerned, of course, essentially with the modification of different human activities in order to optimize environmental conditions, and acts as a kind of community safeguard in order to prevent the maximization by any sectoral interest which could prejudice the use of the environment for the whole community.

Foremost among the problems which the Stockholm Position Paper dealt with was that of world population growth. The Commonwealth, of course, has a critical part to play in resolving this problem, since some of the most severe difficulties in regard to numbers of people occur in Commonwealth countries. I quoted official statistics which predicted a doubling of world population by the end of the century, and that of the estimated total figure of 6,500 million people, over 50% would live in urban areas; and also that people living in settlements above 20,000 people would probably increase from 66% to 81%, i.e. from 600 million to 2,100 million. The population of rural areas in developing countries was expected to grow from 1,910 million to 2,906 million by the end of the century, i.e. an increase of 996 million. In South Asia alone, an important area of Commonwealth interest, it is estimated that another 530 million people would be added to the rural population.

In the light of these trends, the fundamental questions posed in the report were as follows:

a) What should be the world population distribution?
b) What should be the distribution of the productive forces which influence population distribution?
c) What should be the system of settlement networks?

I record this because the planning of human settlements depends absolutely on some kind of ecological balance whereby both urban and non-urban areas can be regarded in a synthetic way. It seems too obvious to state that cities and towns are dependent on rural areas for their food supply. Before the industrial revolution towns were small and agricultural potential large, and in the case of Britain it was the development of that potential which was so significant in enabling the revolution to take place. After the industrial revolution the rapid development of large scale transport enabled food to be carried from almost any agricultural source to any city or town, and an illusion was created that some of the latter could expand indefinitely. Today, however, the loss of good agricultural land, particularly by reason of the rapid spread of human settlements, can be a serious factor in food production, while the problems of mass migration in the developing world caused basically by rapid rural population increase, are preventing any rational environmental solutions. It is essential for everyone to be aware that the good planning and distribution of human settlements is linked directly to the good planning and organization of non-urban land for agriculture, forestry and all the other non-urban uses. Such a statement seems almost too obvious until one realizes that most Commonwealth countries are not yet in a position to declare that it is true, or even that it has been adopted as government policy.

Environmental planning, therefore, enables problems of

human settlements to be considered, first within a global context, then a U.N. Regional (i.e. continental) context, and finally country by country. When seen in this way, human settlements in any one country depend for their basic future welfare on a resolution not only of world population numbers and demographic balance, but of some equalization of economic conditions, for the second most critical problem is to bring into nearer equilibrium the vast disparity in economic levels, so that the GNP per capita will not vary so greatly as it does today between £2073 and £50. Figures of this kind tend to lack significance, and it is very important to go out in the world to see what £50 looks like on the ground.

Among the countries at the bottom of the list those of present or former Commonwealth countries have almost a majority, and they include Uganda, Sudan, Gambia, Pakistan and India. Likewise, near the top of the list one finds Canada, Australia, New Zealand and Britain. Assuming, however, that out of the deliberations of the United Nations will come some kind of world plan (4) which will develop policies for world population, income levels, and the terrestrial distribution of settlement networks, and that within such a terrestrial plan there will be U.N. Regional plans, one can return to plans and policies for each country.

The Position Paper suggested that every country should have a National Environmental Plan based on appropriate environmental policies. Such policies would include amongst many others, a general identification of settlement strategies and networks, together with the setting out of basic per capita environmental standards related to homes, work, recreation and other basic environmental human requirements. Within the Commonwealth, Ghana (5) has already made good progress in this respect. Such National Plans and policies would be of a broad strategic and flexible character, subject, of course, to periodic modification and change, and would be related to appropriate national budgetary forecasts. The latter, incidentally, are often mistakenly called 'Economic Plans', and a great deal of confusion has arisen in the world of semantics ever since the Soviet Union first called its national economic forecast a National Plan. Within the National Environmental Plan there would be organized the appropriate Regional Plans, with the appropriate regional planning machinery. This is, incidentally, being evolved in Scotland (6) at the present time, and the Commonwealth should, therefore, have available a very pertinent experiment in regional planning and policy making in this respect.

Regional plans are automatically very much concerned with the distribution and size of settlements, and one of their most important aspects is to delimit urban areas and urban populations, since it is clear that no city or town is able to perform this task itself. Perhaps most important of all is the development of positive and powerful policies for rural settlements. It is clear that both within and without the Commonwealth this problem remains to be solved. No country has yet produced a satisfactory solution which gives rural dwellers the knowledge, or even the hope, that their lives may be as positive, varied, and viable, as those of urban dwellers. Until this fundamental problem of human settlements is solved not only will rural areas continue to decay, but the larger urban settlements, and particularly the big metropolitan conurbations, will increasingly become uncontrollable in environmental terms.

In 1943 I gave a broadcast (7) on All India Radio in which I stated a few obvious facts about the metropolitan conurbation of Calcutta. These facts, of course, had already been evident to earlier planners like Patrick Geddes, whose reports (8) on Indian cities are still some of the most valuable contributions to planning literature. I pointed out then that Calcutta was already much too large for it ever to provide good human conditions for its citizens, and that a macro-regional plan for Bengal was essential in order to redistribute the population in a more rational and sensible way. It seemed to me at that time that environmental conditions in Calcutta could only get worse if the present trends were to continue. It is not necessary to make a visit to that city today to see how correct the prediction was.

Assuming, however, that each country will develop appropriate regional plans and policies, coupled with powerful and effective implementation procedures, then at the local level it is essential to have urban plans for all human settlements. It is important to remember that in some cases these plans may be concerned with a reduction rather than an increase of size and population of some human settlements, as is so desperately required in the case of

Calcutta. It is, of course, fundamental that planning for human settlements should be based on human environmental needs. Although this is an easy statement to make it is enormously difficult to determine what human needs are in environmental terms, since needs, wants and aspirations all have a way of becoming so mixed up that the planners find great difficulty in arriving at responsive and realistic conclusions, let alone solutions.

The Commonwealth as a whole could, however, look at the problem of basic environmental standards. For instance, what are the minimum spaces to which every person has a right for living, working, and recreation? At the present time a few people have all the space they require and more, but the majority suffer from space deprivation, but just how much it is very difficult to determine. Since almost every type of settlement is available for study within the Commonwealth, from Hong Kong to the suburbs of Toronto, here is a specific and worthwhile task. It was included as one of the recommendations in the Stockholm Position Paper as a task which should be undertaken by the United Nations as a whole, but in the meantime positive action could start within the Commonwealth. This relates, of course, to wider ecological problems concerned with human criteria and standards, such as the appropriate per capita per diem amount of food and water, basic health requirements, etc. and it is obvious that from a CHEC point of view such studies would link across many disciplines.

The Planner's Role

When one looks at this complex network of organizations proposed for the planning of human settlements at national, regional and local levels, it is clear that it is no narrow specialist task. The concept of planning education as originally evolved by such planners as Geddes and Abercrombie was that it should be post-graduate and inter-disciplinary in character. Instead of the single Master Planner there would be a planning team in the form of a composite mind. The planners as applied specialists could each link back to their own basic disciplines where necessary, and thus involve a wide range of experts in the environmental planning process. At each level too, the task of environmental planning would link closely to a whole range of other social and economic activities. An environmental plan means nothing unless it is implemented in the form of dwellings, schools, factories, offices, shops, roads, parks and all the other components of human settlements, and it has to be financially viable and socially acceptable.

The planners are seen, therefore, essentially as members of the broader ecological team with their own expert group of tasks linking closely to all the others. In the case of the planning of human settlements they would generally expect to have the chairman's role in terms of environmental planning, but there are so many other aspects of human settlements which require study that it must always be a team task.

The Role of CHEC in Human Settlements

From the preceding arguments it should be clear that an organization like CHEC could have a very influential role to play in regard to world settlement problems. Certain tasks stand out in terms of priority:

> First, the countries of the Commonwealth, large and small, rich and poor, could get together through the auspices of CHEC in order to wrestle with the difficult but critically important problems of basic human standards. As a planner I would obviously like to see a study of environmental standards high on the list.
>
> The second task would be, of course, to persuade all the Commonwealth countries to adopt such standards and to develop some kind of mutual responsibility so that all could be brought up to a bare minimum. In this respect, it should be remembered that if very large numbers of very small dwellings are built in order to meet the desperate current needs of mass urbanization in developing countries, then this legacy is liable to lie as a great shawdow across the environmental potential of succeeding generations. Only now in the great industrial cities of Great Britain are such legacies handed down from the industrial revolution being resolved, and even now new environmental problems are being created for the citizens of tomorrow in terms of over high densities and over high buildings.

The third priority task could be the giving of advice to member countries in terms of environmental planning organization, so that appropriate and effective planning departments can be set up where necessary at national, regional and local levels. The experience of Commonwealth countries such as Great Britain, where environmental planning at the local level is highly organized; and, in such operations as New Towns, highly effective, could be of considerable benefit, as could the experience gained in Chandigarh in India, Canberra in Australia, Don Mills in Canada, and the planned satellites of Singapore.

Finally, CHEC could have a significant role to play in terms of education for the planning of human settlements. A strong case has already been made for CHEC to become involved in advising member countries in regard to the broader aspects of ecological education, and every year sees new courses in ecology being set up in Commonwealth universities.

For the specific problems of human settlements there is, of course, a paramount need to educate environmental planners and here again new departments of environmental planning are being set up in increasing numbers throughout the Commonwealth. These departments are still new, and particularly because they are inter-disciplinary in character, are subject to many kinds of stresses and strains within academic environments. They need support and help from CHEC and similar organizations in terms of more facilities and finance for student grants for study, travel and research, and all the other normal academic requirements.

In addition to education there is also a need for better institutional machinery. Great Britain again has led the way with its Royal Town Planning Institute, but a great deal of assistance is required in many other Commonwealth countries in order to create strong and effective planning institutes.

I have concluded this paper with a few practical recommendations, and I would only like to remind all those concerned with ecological problems that human settlements are one of the most important aspects of human life. They are the places where the great majority of the people are born, live and die, and unless we can achieve some kind of environmental quality in our human settlements, it is impossible to describe us as a civilized community.

NOTES

(1) UNEP/GC/35 28 January 1975
(2) Summarized for the Commonwealth Universities Congress, Edinburgh 1973, in a paper entitled 'Planning for Human Settlements', July 1973
(3) Human Settlements, No 1 Vol 1, January 1971 et seq. published by UN New York
(4) Two non-governmental agencies have made outstanding contributions. They are (1) The Athens Ekistical Institute, recorded in Ekistics, by C Doxiades, 1968; and (2) The Club of Rome, recorded in The Limits to Growth, 1972, and Mankind at the Turning Point, 1975.
(5) Ghana, A National Physical Development Plan, Accra, 1965
(6) The New Scottish Local Authorities, HMSO 1973
(7) Recorded in An Introduction to Planning, Assam, 1944
(8) Summarized in Patrick Geddes in India, Ed. J Tyrwhitt, London 1947

CONTRIBUTORS

Dr Stephen Boyden
Professorial Fellow
Head of the Urban Biology Group
John Curtin School of Medical Research
Australian National University
P O Box 334
Canberra City
A C T 2601
Australia

The Hon B J Danson
Minister of State for Urban Affairs
House of Commons
Ottawa K1A 0P6
Canada

Professor L O Gertler
Director
School of Urban and Regional Planning
University of Waterloo
University Avenue
Waterloo
Ontario
Canada

The Hon Allan Isaacs
Ministry of Mining and Natural Resources
P O Box 223
Kingston
Jamaica

Professor P E A Johnson-Marshall
Department of Urban Design and Regional Planning
University of Edinburgh
Old College
South Bridge
Edinburgh
Scotland EH8 9YL

F Z Kutena Esq
(Formerly Senior UN Adviser, Bangladesh Government)
C/o 12 Denning Street
Coogee
New South Wales
Australia

Dr Letitia E Obeng
United Nations Environment Programme
P O Box 30552
Nairobi
Kenya

Professor John L Roberts
Department of Political Science and Public Administration
Victoria University of Wellington
P O Box 196
Wellington C 1
New Zealand

Sir Hugh Springer
Secretary-General
Association of Commonwealth Universities
36 Gordon Square
London WC1H OPF
England

EDITORS

J Owen Jones Esq
Director
Commonwealth Bureau of Agricultural Economics
Dartington House
Little Clarendon Street
Oxford
England

Dr Paul Rogers
Department of Applied Chemical and Biological Sciences
The Polytechnic
Queensgate
Huddersfield HD1 3DH
England

INDEX

D

E

F

G

H

I

J

K

L

M

MIX
Papier aus verantwortungsvollen Quellen
Paper from responsible sources
FSC® C105338

If you have any concerns about our products,
you can contact us on
ProductSafety@springernature.com

In case Publisher is established outside the EU,
the EU authorized representative is:
Springer Nature Customer Service Center GmbH
Europaplatz 3, 69115 Heidelberg, Germany

Printed by Libri Plureos GmbH
in Hamburg, Germany